OXFORD TELEVISION STUDIES

General Editors **Charlotte Brunsdon**
John Caughie

Television and New Media Audiences

Television and New Media Audiences

Ellen Seiter

Clarendon Press · Oxford
1999

Oxford University Press, Great Clarendon Street, Oxford OX2 6DP

Oxford New York
Athens Auckland Bangkok Bogotá Buenos Aires Calcutta
Cape Town Chennai Dar es Salaam Delhi Florence Hong Kong Istanbul
Karachi Kuala Lumpur Madrid Melbourne Mexico City Mumbai
Nairobi Paris São Paulo Singapore Taipei Tokyo Toronto Warsaw

and associated companies in
Berlin Ibadan

Oxford is a registered trade mark of Oxford University Press

Published in the United States
by Oxford University Press Inc., New York

First published 1999

British Library Cataloguing in Publication Data
Data available

Library of Congress Cataloging-in-Publication Data
Seiter, Ellen, 1957–
Television and new media audiences / Ellen Seiter.
— (Oxford television studies)
Includes bibliographical references and index.
1. Television viewers—United States—Attitudes. 2. Television viewers—Research—United States. 3. Television viewers—Social aspects—United States. I. Title. II. Series.
PN1992.3.U5S35 1998 302.23′45′0973—dc21 98-28808

ISBN 0-19-871142-5
ISBN 0-19-871141-7 (Pbk.)

10 9 8 7 6 5 4 3 2 1

Typeset by Graphicraft Limited, Hong Kong
Printed in Great Britain
on acid-free paper by
Biddles Ltd, Guildford and King's Lynn

Oxford Television Studies

General Editors
Charlotte Brunsdon and **John Caughie**

OXFORD TELEVISION STUDIES offers international authors—both established and emerging—an opportunity to reflect on particular problems of history, theory, and criticism which are specific to television and which are central to its critical understanding. The perspective of the series will be international, while respecting the peculiarities of the national; it will be historical, without proposing simple histories; and it will be grounded in the analysis of programmes and genres. The series is intended to be foundational without being introductory or routine, facilitating clearly focused critical reflection and engaging a range of debates, topics, and approaches which will offer a basis for the development of television studies.

For **Roy Metcalf**

Acknowledgements

I RECEIVED MANY good ideas and kind words of encouragement from Cynthia Chris, Susan Davis, Lynn Hudson, Kathy Krendl, Rosemary Morrison, Dan Schiller, Jane Rhodes, Karen Riggs, and Kathleen Rowe. My thinking about television has been shaped by friendships and intellectual exchanges with Ien Ang, John Caldwell, Henry Jenkins, Chuck Kleinhans, Julia Lesage, David Morley, and Lynn Spigel.

Field research was funded by the Indiana University College of Arts and Sciences and the Center for the Study of Women in Society at the University of Oregon. Without the enthusiastic participation of the parents, pre-school teachers and daycare-givers in my interviews, this book would not have been possible.

I am grateful to Charlotte Brunsdon for her friendship over many years, and especially for her tactful and incisive comments on this book.

Joseph, Henry, and Anne Metcalf have proven to be my best audience. Thanks to each of them for enthusiastic conversations with me about television, for their eager interest and pride in my research, and for their willingness to play resourcefully in offices, corridors, and on campus lawns while waiting for me to finish up work.

Ellen Seiter
San Diego, California

Contents

1

Introduction

THE SCENE IS a classroom of four-year olds at an upper-middle-class nursery school in a US Midwestern suburb. About twenty children are present, fifteen of them boys. Two teachers are present, one is a woman in her late fifties, the other is a student teacher in her early twenties. It is late morning clean-up time, when the teachers attempt to secure the children's efforts to tidy up the classroom before the children go to the outdoor playground for recess.

Two boys are playing in a corner of the room with tiny toy cars. One is a slender, white, extremely talkative boy named Ian. The other is a small, Chinese-American boy named Wu. Bedlam is all around them.

A third boy, larger and older than Ian and Wu, approaches. His name is Michael. 'Can I play with you?' he asks.

'Sorry, but me and Wu are playing', Ian replies.

A few minutes later, a fourth boy, Casey, who is even larger than Michael and very rambunctious, joins them in play without asking permission.

While Casey plays with the cars and blocks he sings, 'Flintstones, meet the Flintstones have a yabba dabba doo time . . .'

Ian and Wu are silent. After a pause in the singing, Ian strikes up some conversation:

IAN: Guess what? You know I heard that the Flintstones are going to see the Jetsons.

CASEY: You mean on the cartoons?

IAN: No, the show.

CASEY: On the Flintstones show they're going over to see the Jetsons.

The conversation fades out here and the boys continue playing until the student teacher approaches. She asks, 'What are you doing?' in an accusatory way that implies that they should be putting the toys away rather than playing with them.

In high-pitched, joking voices, the boys reply, 'We're trying to sort these [the blocks] out.' Lingering for a moment to check up on them, the teacher observes them hiding the tiny cars behind the blocks as they clean up. Clearly irritated with them for breaking a frequently repeated rule about returning toys to their rightful storage containers, the teacher switches to a commanding tone: 'I want Casey to put the cars away, Wu to put the big blocks away, and Ian to put the small blocks away.'

As she walks away, the boys erupt in laughter, exhilarated by their naughtiness and the discovery of their crime. Casey begins to sing again, and the other boys join in, singing loudly: 'Flintstones, meet the Flintstones.' As they sing, they grow more raucous, boisterous. Wu is laughing hard, intensely enjoying his inclusion in the singing of the song and the rebellion it signifies. The boys repeat the same song lyrics over and over again: 'Flintstones, meet the Flintstones.'

Michael, curious about the good time the three boys are having, wanders over and tries to join them again. Casey immediately grabs a plastic pan and rhythmically hits Michael over the head with it in time to the beat of the music, while singing 'Have a yabba dabba doo time, a dabba doo time . . .'

As a qualitative audience researcher observing this scene after some twenty hours of visits to this classroom, and nine months of visits to the school, I see in the boys' classroom interaction the complex ways that popular television is embedded in interpersonal communications, in gendered conflicts, and in the exchange of tokens of cultural capital. To the student teacher walking over to break up the rough-housing between Casey and Michael, the scene might confirm the widely shared conviction among teachers that television produces violent behaviour in children and causes disruptions in the classroom. It must be admitted from the outset that qualitative audience research can do little to confirm or deny such a hypothesis about television's effects. Qualitative research can, however, offer an interpretation of this scene that takes account of the contextual factors at work here, and the various uses of television as a form and topic of communication with others in social settings. The primary contribution of ethnographic audience research since the 1970s has been its demonstration that media consumption is embedded in the routines, rituals, and institutions—both public and domestic—of everyday life. The meanings of the media, whether in the form of print, broadcast, or recorded video, or computer forms, are inseparable from and negotiated within these contexts.

Media consumption and interpersonal relationships are closely intertwined. It is erroneous to treat media communication as existing separate from or simply interfering with interpersonal communication. For example, family members may use the media as an occasion for conversation and affection (Palmer 1986; Lull 1991), while office workers may use television as a means of facilitating chat (Hobson 1989; Seiter 1996). Alternatively, engagement with the media may form a message about unavailability, especially in domestic space, as when reading is used as a buffer against conversation or demands (Bausinger 1984; Radway 1984).

In this classroom scene, Ian and Casey, two boys who play together infrequently at school, use chat about *The Flintstones* and *The Jetsons* to make conversation. Later, when the boys join together singing the *Flintstones* theme song as a rebel call, Wu is especially exhilarated by it because he is often excluded by Casey from play—as are the other Asian children. On numerous occasions I have observed Wu strategically deploying his knowledge about toys and TV superheroes to gain entry into play situations with some of the dominant (and white) boys in the class.

References to the media in conversation are complex, and must be understood in context. Conversation about a given media programme or newspaper or radio show does not in and of itself suggest a strong affection for a given media event. Often, conversation about the media is used as a pretence to talk about interpersonal relationships, longings, and desires, or 'taboo' topics such as sexual violence, gay and lesbian sexuality, racial tensions. Media talk is crucially related to the management of social relationships, engaged in as a means of maintaining social connections as much as it is motivated by interest in the media *per se*. The discussion of sports and soap opera in the communal areas of the workplace—the coffee machine, the lounge, the cafeteria—is one example of this. Media references may be used inventively, 'poached' by media consumers to suggest a meaning quite different from those intended by producers (Jenkins 1992). *Star Trek* fans rewrite the character relationships and plots of the original television programmes to express interests never actually covered by them (Bacon-Smith 1992; Jenkins 1992).

It would be erroneous to infer that these three boys are avid fans of *The Flintstones*, for example. Although Michael, who initiates the reference to the cartoon, might recently have seen either the television or the live-action film version, released on video at about the time of this conversation, *The Flintstones* was not a frequent topic of his chat. Something about the blocks and the cars seemed to remind him of the Stone Age cars and garages on the cartoon—or perhaps the song had simply stuck in his mind. Rather than interpreting this scene as an instance of direct effects, it seems typical of the casual references, remembered jingles, and the like that children reuse for their own purposes (Buckingham 1993, 1996). In this case, the incongruous introduction of TV references into the classroom seems to work for the children and the teacher as a battle cry of rebellion. One of the most salient messages of this use of the *Flintstones* song is: I am doing something naughty, and I refuse to clean up.

Dynamics around media consumption are gendered in terms of space (who uses what media where), time (who has more leisure time), allocation of resources (who spends money on the media), and the use of media technologies in the home (who gets priority in using the TV, the telephone, the computer—or in reading the newspaper). Domestic studies frequently document the polarized worldviews—and diametrically opposed media tastes—of some men and women. Preferences for television, for example, tend to fall into gender patterns and have symbolic significance as such. Audience researchers have frequently noted battles in the living-room stemming from women's boredom at sports and distaste for action-adventure films. On the other hand, some men are derisive of soap operas and weepies (Morley 1986; A. Gray 1987; Press 1991). Among young children, mothers and female teachers often battle to keep little boys away from violent cartoons, such as *Mighty Morphin Power Rangers* or forbidden video games, such as *Street Fighter*. Many women disapprove of the forms of masculinity celebrated in action-adventure genres (Seiter 1993). Housework and leisure are often differentially distributed within households. Ann Gray (1987) found that some of her informants refused to operate the video cassette recorder because they did not want to be assigned yet another chore by family members—that of recording their favourite programmes.

In our classroom example, the pre-school teacher discovers the boys breaking the classroom rules (playing instead of tidying up; hiding toys rather than putting them in the proper storage bin; rough-housing), and the rule-breaking is associated with children's talk about TV at school. Ian, Wu, and Michael are enjoying a moment of male bonding through TV references while simultaneously insisting on their right to play instead of work, or to ignore the teacher's orders, or to leave the cleaning up to the girls. Pre-school teachers are almost always female—membership in NAEYC (the National Association for the Education of Young Children) is over 90 per cent female, for example—and teachers in this classroom (and many others) reported devoting most of their time to disciplining boys. Discipline problems and references to popular culture are closely linked in the teachers' minds, especially with action-adventure programmes such as *X-Men* or *Power Rangers*—forms of television that many teachers do not approve of and forms of (superheroic) masculinity they do not find attractive.

Television is the least legitimate of media forms. Reading any form of material is usually granted higher status. When people talk about television with academic researchers, the discussion is often inflected by this illegitimate status, prompting frequent apologies or an underestimation of TV viewing. Many studies of middle-class adults have documented this defensiveness, especially among women (Radway 1984; Seiter 1990; Press 1991). Ethnographic studies have usefully pointed out the ways in which such hierarchies produce defensiveness about certain forms of media consumption. This defensiveness may lead to what sociologists call the 'halo effect' or social desirability factor, manifesting itself in this case as the underestimation of some forms of media consumption (such as television viewing) and the overestimation of other forms of media consumption (such as book and newspaper reading or computer usage). We need to understand each media form (both the technology and the genre) and the segments of the population which express tastes for each form in terms of their places in the cultural hierarchy.

For example, certain television genres, such as public affairs programming and historical dramas, are associated with an educated middle-class audience, who resist being characterized as heavy or even regular viewers, since listening to recorded music, reading novels, or, better yet, leaving the home to go out to attend live theatre, are considered to be more worthwhile and 'improving' activities. Other genres, such as soap operas and situation comedies, are associated with working-class viewers, who may have less cash available to purchase other kinds of cultural good or less freedom to leave the home, may enjoy discussing such popular programmes with others at work, and may openly resist the concept of watching television for education and self-improvement.

Children offer an interesting test case because they are oblivious to many adult notions of cultural prestige, yet have begun to appreciate some of the distinctions between adult and child attitudes towards TV. As early as the age of four, children can appreciate that *The Flintstones* is not normally part of their school's curriculum—not the sort of video title (like a nature documentary or *Sesame Street* episode) that would be approved for classroom viewing. Ian, Wu, and Michael have been taught by their teachers—both directly and indirectly—that they are not supposed to talk about TV in this school, where books are valued, where tapes are rarely shown, where show-and-tell objects are censored. TV takes its place in the repertoire of forbidden references, like those to smelly feet or body parts or diapers. In fact, TV songs or jingles are often sung moments before or after crude language or jokes are voiced. No wonder many teachers hate popular children's TV, when it is associated with bedlam, rule-breaking, forbidden activities.

If I had visited this classroom on a single day, I would have had no possibility of understanding the ways in which media references do and also do not serve young children's negotiations and friendships. It is significant, for example, that on many days when I visited the class no references to television were made within my hearing. It is equally significant that fights such as the one between Casey and Michael occurred very frequently, sometimes linked to the media, as in enacting *Power Ranger* karate kicks, and sometimes not. During my first days of classroom observations, the boys were curious, wary, and guarded around me. It was only after several weeks that Casey announced, after starting a fight nearby and looking at me, 'She's not a teacher!' and that the boys engaged in illicit behaviour within my hearing. Thus, doing ethnographic audience research necessitates making contact with informants repeatedly, for as much time as possible, and under as many different circumstances as possible.

This book explores television consumption in the contexts of the home and the classroom, as mediated by family relationships and the relationships between teachers and children. I will discuss illustrative cases from field projects that I have conducted over the last decade, focusing on a particular set of issues that have preoccupied me: gender (both in preferences for particular genres and in differential patterns of access to television sets and computers in the home); adult–child relationships—both in terms of the pervasiveness of television in young children's lives and as a recurring subject of struggle and negotiation; and, finally, the leisure gap between men and women, how domestic chores and the labour of childcare are unequally distributed in the home, and the effect of this on opportunities for using television for entertainment.

A considerable selection process is at work here, both in terms of reporting the data from these studies (each of which could easily fill an entire book) and in terms of references to the now considerable field of TV studies: theoretical work, textual criticism, and empirical audience research. Throughout the book I offer a very selective overview of model case studies and of theoretical issues: the role of language, the othering of the subject, the foundational work on media consumption in the domestic sphere, the influence of French sociologist Pierre Bourdieu's work on concepts of class, and the postmodern critique of ethnography. Excellent books by Shaun Moores (1993), David Morley (1992), Virginia Nightingale (1996), and Roger Silverstone (1994) offer a comprehensive account of the development of work on audiences. This volume does not attempt to do so, or to offer a historical account of the development of cultural studies and TV studies. Rather, I have focused only on those studies that have motivated my own interest in conducting this form of research or which suggest fruitful directions for future research. Much excellent work has been left out, because it employed methods or dealt with kinds of research subject, television genre, or location of viewing that fell outside my immediate areas of interest.

Most of my examples will be drawn from two field projects I have conducted: a participant ethnography of a parents' support group and an intensive interviewing project with pre-school teachers and daycare-givers. The studies were undertaken between 1987 and 1995 in small college towns in the Pacific Northwest and Midwestern regions of the USA. My purpose in presenting examples from these projects is twofold: to examine the benefits and limitations of specific methods and types of research design for studying audiences, and to emphasize the importance of continuing to connect audience research to questions about gender relations, the domestic division of labour, the leisure gap between men and women, and social class as a determinant of TV viewing and feelings about TV viewing. Throughout the book, I explore how people's informal theories, or 'lay theories' of media effects—strongly influenced by advice literature and popular discourse about appropriate media consumption—constantly inform practices of media consumption and mask status distinctions maintained through TV knowledge.

In Chapter 2, I review some of the relevant literature on television audience research, addressing the particular challenges of importing ethnographic method from anthropology to the study of a phenomenon as pervasive and yet transient as television viewing. After a brief comparison of approaches to the study of audiences common to the US mass communications research tradition, I summarize a few influential studies informed by the cultural studies tradition. Finally, I survey some of the issues in research design, such as interviewing versus observation, the length of contact required, the analysis of linguistic forms, and the practice of self-reflexivity, which seem to me essential to doing good fieldwork.

In Chapter 3, I present examples from a longitudinal study of a parents' support group in which I was a participant. In this study, I wanted to show the trickle-down effect of the academic representation of children's (and women's) viewing as passive and, therefore, bad, and its consequences for the labour of middle-class mothers. I collected data formally through group interviews about television and advertising, but also informally through monthly conversations with the sixteen members of the group about the media, and about their worries over television's impact on children. Advice literature in women's magazines and educational handouts from pediatricians and teachers incessantly urge parents—and this, of course, means mothers—to monitor and control their children's television viewing. How do mothers make these determinations about what is acceptable for children to watch and what is not? How are these decisions and rules shaped by class background, educational aspirations, and mothers' own struggles with the combined load of housework, childcare, and, very often, paid work outside the home? How much are they influenced by a particular ideology of childhood embedded in developmental psychology? What is the relationship between adult women's viewing habits and the television viewing of their children?

Chapters 4 and 5 present case studies from research with twenty-five pre-school teachers and daycare-givers. The interviews took place in 1994 and 1995 and were carried out by me and my research assistants Madelyn Ritrosky-Winslow and Karen Riggs at the subjects' places of employment, usually on a lunch-hour or long break. At the end of the interviews, subjects were asked to keep a one-week diary of all media used both in the classroom and in their leisure time. Subjects were interviewed again a week later, and were paid $100 for their time and expertise. Interviews focused on the extent to which television is talked about with children by their care-givers outside the home. I attempted to record the range of praise and criticism for children's television among childcare professionals—and the schemes of value at work in judging some videos as unacceptable for classroom use and others as acceptable. I also gathered information on teachers' leisure-time media preferences and their own childhood experiences with television in order to correlate these with perceptions about children's television viewing.

In Chapter 4, I present four case studies to illustrate how beliefs about media effects are implicated in fears about the future, the degree and nature of social aspiration, the moral judgement of popular media, and consumer culture. This study suggests ways that negative representations of children's television viewing find their way into the educational system in real ways: punishing children who know too much about TV; rewarding those who show an early familiarity with computers. A studied, conspicuous ignorance about television is the mark of distinction, and like all distinctions it is valued because it is so difficult to maintain. This research also argues for the importance of moving out of the domestic context to examine the meanings of television in the workplace and in institutional settings. Rather than abandon the project of charting the sexual politics of domestic leisure and media consumption, I think the study of TV consumption on the job and at home can clarify the especially problematic position of women as television viewers, as gatekeepers for the family, and as exploited workers.

Chapter 5 examines interviews with Christian fundamentalists who were part of the sample of pre-school teachers. In these case studies a quite different set of concerns emerge: Satanism, depictions of supernatural powers, gay and lesbian relationships, subversions of parental authority. This chapter explores the substance of the fundamentalist critique of children's media culture. How does this parallel as well as deviate from the traditional liberal critique of children's TV, or the restrictions advocated by secular authorities, such as educational psychologists? How do fundamentalist women view their work with small children and what is the significance of their gatekeeping role?

Chapter 6 suggests that television audience research should engage with the study of computer use. The integration of research on television with research on computer use makes sense from the perspective of communication theory, the sociology of everyday life, and the political economics of the media industries. Ethnographic approaches seem especially well suited to the study of communication technologies at home and at work. Television research has much to contribute to an understanding of the Internet, which is replicating many of the generic forms and practices of audience segmentation familiar from the television industry. On the other hand, bringing computers into the mix of media studied by audience researchers will help to focus attention on questions about media consumption outside the living-room.

Each chapter focuses on case studies of specific kinds of experience with the media: as parents of young children, as teachers and care-givers, and as members of Christian fundamentalist churches. But this is not exclusively a book about children or about women in their roles as gatekeepers of children's media consumption. Rather, I try to explore the contradictions that arise among the multiple roles that each media consumer adopts in everyday life: teachers are often also parents; fundamentalists are also workers; adults have powerful memories of childhood experiences with the media; and computer aficionados are also often television viewers.

This book uses empirical case studies to develop a fuller picture of the contradictions inherent in the relation of television to the lives of families with small children, to the dynamics surrounding television viewing in classrooms, and to the parallels between gendered relations to television and gendered relations to computers. These admittedly selected interests among the possible range of topics which audience studies can address grow out of my teaching in both media and women's studies. My goals in this book are to acquaint readers with the range of theoretical and methodological issues that inform audience research, to offer advice and discussion to the reader who contemplates carrying out such a project, and to encourage a high degree of self-reflexivity around issues of gender and class. Pierre Bourdieu argues that nothing is better able to express social differences than the field of cultural goods—and here I would substitute television—because 'the relationship of distinction is objectively inscribed within it, and is reactivated, intentionally or not, in each act of consumption, through the instruments of economic and cultural appropriation which it requires' (1984: 226). The challenge for audience researchers is to investigate popular tastes and explain how these tastes are distributed in relations of domination. To do so also necessitates recognizing the dominance of academic interests within the system of cultural distinctions.

2

Qualitative Audience Research

LOOKING AT MEDIA in the context of everyday life presents many research problems. How can we study the way viewers interpret television programmes in routinely occurring settings, such as the home or the school? What is the best way to get people to talk about the meanings they derive from television programmes when they may be unaccustomed to interpreting TV material explicitly? How can we study what goes on when people consume media, when so much media consumption takes place in private, and in the domestic sphere, in the context of intimate relationships? How can we study conversation about the media, short of trailing a person throughout the day? How does the researcher herself influence, inhibit, and change the ways people will talk about the media? The media overlap with many dimensions of social life, such as gender roles in the family, political beliefs, social networks of kin and friendship, routines of the clock at home, work, and school, allocation of household resources, and the organization of the workplace. How do we draw the line in our data collection between audience research and the study of society, the family, the community? Should we draw such a line? These are some of the problems facing researchers interested in using qualitative methods to study media audiences.

In this book, I wish to argue for the usefulness of ethnographic methods in studying television viewing and computer use. The dozen studies that I will review in this chapter share a use of semi-structured and open-ended interviews. Most commonly used is a procedure of interviews, often with groups of subjects, where the interviewer follows an outline of interview topics and questions, but allows informants to raise topics not included on the list. Still, considerable variation exists from study to study in terms of contact time, the role of the interviewer, the adherence to questions written out in advance, the settings for the interviews, the use of group interviews, and the means of contacting subjects.

I begin, then, with a consideration of ethnographic method, the anthropological tradition in which its research procedures were developed, and the difficulties in translating this model to the study of media in contemporary social life. I compare the tradition of US mass communication research based on a media effects or uses and gratifications paradigm with audience studies influenced by a cultural studies paradigm. Next, I discuss examples of two kinds of study: one is based on Stuart Hall's encoding–decoding model and focuses on viewer interpretations of specific television programmes; the other is based more broadly on the domestic contexts of media consumption and the way these are structured by family relationships. The next section takes up the influence of Pierre Bourdieu's concepts of cultural capital and aesthetic dispositions on the field of television studies and on my own research, which is presented in the following three chapters. Finally, I discuss the importance of theories of language to audience research, and the importance of self-reflexivity about the researcher's role in shaping the interview process and the 'othering' of research subjects.

Ethnographic Method

Ethnographic method is a distinctive research process developed within anthropology and sociology involving extended periods of participant observation and emphasizing descriptive writing of both field notes and the final ethnography. Ethnography's goal is to produce a holistic description of a culture. As anthropologists George Marcus and Michael Fischer describe it: 'Ethnography is a research process in which the anthropologist closely observes, records, and engages in the daily life of another culture—an experience labeled as the fieldwork method—and then writes accounts of this culture, emphasizing descriptive detail' (1986: 18). Very few media audience studies, even those using ethnographic or qualitative methods, have measured up to the normative standards of ethnography proper. Most of the time, 'ethnographic' has been used very loosely to indicate any research that uses qualitative interviewing techniques. Many of the most influential audience research projects, such as David Morley's study of lower-middle-class London families (1986), Janice Radway's work on middle-aged readers of paperback romances (1984), Ien Ang's analysis of letters from *Dallas* fans (1985), Ann Gray's study of video cassette recorder use (1987, 1992), and Elihu Katz and Tamar Liebes's cross-cultural study of focus groups discussing *Dallas* (Liebes 1990), were not designated ethnography by the original authors, but were labelled ethnographic in secondary accounts. While ethnographies are based on long-term and in-depth fieldwork, most audience research has been based on brief periods of contact, in some cases less than one hour, with the informants. Also, while ethnographic methods have traditionally been used to study culture as a whole, media researchers study only one aspect of a culture—such as television—when using this method, and attempt to relate it to social identity (Seiter *et al.* 1989: 227).

Some media research does meet the requirements of ethnography, including Marie Gillespie's study of Punjabi youth in Southall, England (1995), Camille Bacon-Smith's account of US *Star Trek* fans (1992), and Angela McRobbie's study of teenage girls at a Birmingham youth club (1991). The difference between these studies and other audience research is that they involved extended contact time over a period of years, and a combination of methods, including quantitative ones. As Gillespie puts it, ethnographic fieldwork 'is characterised by a multiplicity of data-gathering strategies, in a variety of contexts, drawing upon the experiences of a wide range of people over a long period of time' (1995: 60).

Gillespie's work on media use by Punjabi youth in Southall comes closer than any other work on media audiences to satisfying the requirements of ethnography. Much of her study focuses on TV talk. How did these teenagers use British soaps, news, and advertising to negotiate questions of their own identities as South Asian immigrants to London? How do they use videos of popular films produced in India? One segment of the study looks at videos of religious tales (such as the Hindu devotional texts *Mahabharata*) and radio news broadcast in Indian languages, both of which are used by the teenagers to increase their fluency in Indian languages, establish adult identity within the family, and earn the trust and recognition of their elders.

Before embarking on the study, Gillespie had lived in Southall for many years, and originally taught English as a foreign language. Her research included informal work (befriending neighbours, establishing rapport) and a formal role; she taught in a local high school for seven years. As a teacher, she was privy to casual conversations, befriended students, and at times used class assignments to facilitate discussion of the media in everyday life and in the construction of cultural identities. In addition to the ethnographic research she carried out in her formal role at the high school, she became closely involved with a number of families, establishing 'close and reciprocal relations' with them that included making regular visits and attending formal celebrations and casual meals, outings, etc. Gillespie kept audio tapes, video records, field notes, and a field diary (1995: 62).

Gillespie compared her own experiences as the daughter of Irish immigrants to England to the experiences of Punjabi youth, in terms of close-knit family relations, the prominence of religion, and the longing for an idealized homeland (1995: 49–51). However, her study is not especially self-reflexive. She trusts in the classic ethnographic rapport, followed by the necessity of withdrawal from the situation during the writing of the ethnography. Her most important contribution is an understanding of the meanings of the media for diasporic cultures, how these are interwoven with generational conflicts, how language differences are negotiated intergenerationally and through the media, as children become translators of English for their parents, for example, and as children use imported media to enhance their learning of their parents' first language. TV plays a central role in the creation and negotiation of identities in a postmodern world, one characterized by large-scale immigration and travel. 'In Southall the redefinition of ethnicity is enacted in young people's reception and appropriation of TV' (1995: 22). Gillespie thus argues persuasively that media ethnography may be placed usefully at the core of studies of immigration and diasporic communities and postmodern identity formations.

Mass Communications vs. Cultural Studies

Central to the renewed interest in qualitative research on media audiences have been questions of how specific audiences make meanings in their engagement with media in the context of everyday life, an emphasis on audience activity rather than passivity, and an interest in why the media are pleasurable. This move stems in part from the increased agency attributed to the media consumer in uses and gratifications research, a paradigm that succeeded in altering the way media effects were discussed before the 1960s in US mass communications research. As communications scholar Carl Bybee characterized the effects tradition: 'The history of mass communication effects research in the United States is the history of a relentless, empirical search, first for direct, powerful short-term attitudinal effects, and later for the intervening variables which could be regarded as either facilitative or obstructive of those effects' (1987: 195). The uses and gratifications research represented a shift to a more optimistic and less harmful characterization of the relationship between media and audiences, emphasizing active engagement and the ways the media could be employed by individuals to satisfy needs and accomplish personal goals. According to Bybee, uses and gratifications research does not represent a dramatic break from the traditional effects perspective, as it kept intact 'its conservative bias regarding the process by which political power is distributed in society' (1987: 194). The shift from the effects model to uses and gratifications is only an evolution at both the systems and individual levels, as the focus of attention changed from the communicator to the audience. What was left behind was essentially an untenable stimulus-response conception of the media effects process. What was carried along was essentially a limited conception of the media effects process, a lack of explicit social theoretical referents for the interpretation of individual level results, and a consumerist frame of reference (Bybee 1987: 196).

The other influence on audience research has been British cultural studies. The twin influences of the uses and gratifications model and British cultural studies are not easily distinguished in a single piece of research, and often scholars such as David Morley recognize both paradigms as influential on their work. Cultural studies brought to audience research the emphasis on processes of decoding cultural texts, and theories derived from semiotics (Eco 1976) and reader-response literature (Iser 1978). The British cultural studies tradition grew out of ethnographic research carried out at the Centre for Contemporary Cultural Studies (CCCS) in the 1970s at the University of Birmingham, and captured a more nuanced sense of the complexity of television as a text, as well as a conception of audience activity that was informed by Marxist theories of ideology—which brought explicit questions about social power to the research, distinguishing it markedly from the uses and gratifications perspective. The uses and gratifications model is based on a pluralist conception of society—in which there is something for everyone in the media forms on offer—and a functionalist sociological model—focusing on the explanation of social stability. Uses and gratifications research thus lacks a concern with the power relationships that determine both audiences and forms of media production, while the cultural studies model has tried explicitly to address the question of social power on a number of fronts.

The borrowing of ethnographic research methods from anthropology was motivated by a critique of experimental and survey audience research in the mass communications tradition. The critique associated with 'critical communications scholars' addressed research procedures, theoretical underpinnings (especially behaviourism), and institutional influences (such as the preference for quantitative findings). Such research failed to address important questions of reception and audience activity. The charges were that mass communications audience researchers were wedded to methodologies that restricted them to questions answerable through quantitative methods. In particular, there had been too much emphasis on observable behaviours, rather than structures of meaning. This emphasis on quantifiable phenomenon locked mass communications researchers into a cycle of number-crunching. Funding agencies increasingly demanded statistical results. Such administrative research thus followed the norms of market research, where sponsors require clear-cut findings.

Mass communications researchers avoided studying the media in context, preferring instead sanitized, controllable situations (laboratory, telephone interviews), producing data that was irrelevant to everyday life. Often researchers remained ignorant of the media forms they studied and handled media content awkwardly, if at all. Finally, mass communications research lacked a theoretical perspective on language as discourse. This led to a preference for reducing answers to easily codified categories or taking subjects' answers at face value. Similarly, content was reduced to verbal summaries of observable events on screen.

The research that I will describe in this chapter represented a departure from these norms, and bears the influence of ethnographic method. First of all, the new audience studies differ methodologically from quantitative research in that their projects tend to proceed without a clear-cut hypothesis, and investigate multiple research questions interpretively. My research deviates, then, from the scientific model widely adapted by US social scientists (even those working in the uses and gratifications tradition)—with the exception of anthropologists—in the twentieth century. Sample sizes tend to be much smaller than those required for survey research, nearly always involving fewer than one hundred research subjects, in some cases fewer than thirty. Thus, statistical generalizability is sacrificed; the model for such research is the case study, rather than the survey.

In many of the television audience studies I will describe here, quantification is avoided, or relegated to an appendix. Instead, extensive quotation of informants is presented. There is as much interest in the thoughts and feelings of audience members as in their behaviour. Typically, the research requires the establishment of rapport between the researcher and the subject. This may range from conducting interviews in a friendly, open manner to establishing personal friendships with informants.

Traditional social scientists have faulted this work for lack of generalizability, bias, political axe-grinding, failure to employ multiple methods, and a casual and sloppy approach to data collection. Some of this new research was carried out by scholars trained in the humanities, often by European rather than US academics, or by those trained in disciplines influenced by recent European theory (such as semiotics and psychoanalysis), especially literary criticism, textual analysis, and film theory. The CCCS, under Stuart Hall's guidance, was a crucial influence on the development of this work. Like much of the research conducted at the centre, these researchers' work was rooted in Marxist and feminist theory, and questions of class and gender have therefore been central. By contrast, audience researchers in the mass communications tradition have been trained in social science disciplines, especially social psychology, or in journalism (Delia 1987). In the mass communications tradition, especially in the USA, there has been less emphasis on theoretical developments. James Curran, in criticizing the 'new revisionism' in qualitative audience research, has expressed understandable irritation with the failure of some researchers to recognize when they are revisiting questions that have been debated since the 1950s, without reference to any work done before 1970 (Curran 1996: 264–7). Thus the conflict between so-called 'ethnographic' audience researchers and mass communications researchers involves a generation gap, a disciplinary split, and a continental divide.

Encoding/Decoding

A good way to see the differences between a US mass communications perspective on audiences, and a European media studies approach, is in Hall's influential encoding–decoding model and its application in David Morley and Charlotte Brunsdon's work on the news magazine programme *Nationwide*. The centrality of ideology, the variability of interpretation of television, and the complex variables in viewers' interpretations form the core of the theory.

David Morley's first study followed a detailed analysis with Charlotte Brunsdon of *Nationwide*, which was published as *Everyday Television: 'Nationwide'* (Brunsdon and Morley 1978). In the second project, published as *The 'Nationwide' Audience* (Morley 1980), Morley 'explored how that programme material was interpreted by individuals from different social backgrounds, with a view to establishing the role of cultural frameworks in determining individual interpretations of the programmes in question . . . [and] some of the relations between socio-demographic factors (such as age, sex, race, class) and differential interpretation of the same programme material' (1992: 75).

The *Nationwide* study was an attempt to elucidate the encoding–decoding model, adapted from Frank Parkins's work. Discussions of the project frequently neglect the close textual analysis that preceded the audience study. It is crucial, however, to the project's conception, that the research was designed first to elaborate the encoding of the programme before exploring the variety of decodings.

The encoding–decoding model was an attempt to get away from a linear sender–message–receiver model of mass communication. It posits three distinctive types of interpretations or decodings. The *dominant reading* is performed by viewers who accept the programme and its genre completely. These viewers would agree with the dominant ideology (the preferred reading) of the programme without formulating any objections in their minds. Such a viewer uses ideology to explain her own life and behaviour, and her social experiences. In a *negotiated reading*, the viewer inflects his interpretation on the basis of a particular social experience. The viewer may enjoy a 'pick and choose' relationship to the genre, ignoring more disagreeable sections and concentrating on those more to taste. Another way to think of this is in terms of 'shifting' the text slightly to fit individual interests. Here, the media consumer is mainly in line with dominant ideology, but needs to adjust certain aspects to fit her local situation. She might ignore some parts of the show, while focusing on others, providing explanations of events portrayed that suit her own worldview, not all of which may be as strongly 'there' as others. The most radical viewing position is that of an *oppositional reading*—in which the viewer goes against the preferred reading. This type of reading is characterized by annoyance rather than pleasure—as when the reader, recognizing the political motivation of a news programme, says, 'There they go, up to their old tricks again!' (Fiske 1992: 292–8).

The encoding–decoding model insists on the struggle involved in gaining people's agreement with ideology; both because television is complex in how it tells stories, and because how people read television will necessarily be based on their own experiences—what kind of jobs they have, where they live, their educational backgrounds, memberships in unions or political parties, as well as gender, ethnicity, sexual orientation, and class.

For the audience project, Morley gathered focus groups of adult education students who viewed a tape of the programme and then participated in a discussion. While some of the decodings were predictable, given the class position of his subjects, Morley concluded that a more complex model was necessary to determine the ways that social position might predispose TV viewers to make certain types of ideological reading. Contradictory in nature are the responses which individuals may make to different types of programme: audience members may read one programme subversively, another according to a dominant reading; or they may read the same material differently depending on the context. A single individual would not read all of the media the same way: some shows might be laughed off, while others are despised; still other shows might be found to be very absorbing. Interpretations or decodings will also vary.

Morley's *The 'Nationwide' Audience* is a model for John Corner, Kay Richardson, and Natalie Fenton's decoding study, published in 1990 as *Nuclear Reactions*, and explicitly set out as a development of the *Nationwide* study, which it describes as faltering a bit under 'a sometimes vulnerably mechanistic idea of interpretation as "decoding"' (Corner *et al.* 1990: 47). Like Brunsdon and Morley's 1978 study, *Nuclear Reactions* begins with a close textual analysis. In the introduction, the authors warn that the turn to studying television audiences is not a simple return to fieldwork or to empirical investigation in media studies. Rather, it represents 'an attention to the detail of significatory form (image and language) and to the "creative" processes of interpretation of a kind not generally observable in the mainstream social science tradition' (1990: 47). One aim of the work is to reorient reception studies towards the complexities of decoding non-fictional material, complexities that the authors feel were overshadowed by the flurry of interest and research on soap opera reception. Corner *et al.* analyse four non-fiction television programmes on the subject of nuclear power along with the production background of each programme, so as to consider 'matters of producer intentions and production contexts' (1990: 27). *Nuclear Reactions* examines the 'textualization' of nuclear energy as a political issue in a dramatization, a promotional tape by the energy industry, and two examples of investigative reporting—where '*processes* of selective transformation at work as evidence, ideas, images and speakers from the world anterior to television were processed and assembled within different, self-contained rhetorical systems' (1990: 10; emphasis in original).

In this project, the authors sought to form natural groups (the respondents were already known to each other through some other circumstances) based on political affiliations (Labour, Conservative, Friends of the Earth), workplace relationships (nuclear plant workers), or school cohorts (medical students, high school students). Subjects were shown a taped programme on the topic of nuclear power, and discussions ensued. The authors describe their approach to the discussions as 'ethnodiscursive', to suggest the ways that samples were from natural groups and 'close attention . . . was paid to the language used by the respondents in articulating and negotiating their responses' (1990: 50). The analysed transcripts yielded a number of frames—political, civic, personal, evidential. The authors argue that 'respondents appropriate the programmes, or aspects of them, from within particular frameworks of understanding, which supply them with criteria of evaluation both for programme forms and contents' (1990: 50).

Corner, Richardson, and Fenton caution in their conclusion that they are not delivering 'findings' from this study in the conventional sense. Rather, they extract suggestions about the TV process from comparisons among different groups of subject (often holding different positions relative to the public issue of nuclear power) and from their analysis of particular frames. Despite their refusal to report traditional findings, they were struck by:

> *the extensive presence in viewers' accounts of the 'civic' frame*, a frame which strenuously, and sometimes with great difficulty, seeks for overall 'fairness' above the weighted presentation of even a preferred viewpoint . . . It is the single most powerful regulator of interpretative assessments we found and it frequently provides the parameters within which a critical scrutiny of *forms* is carried out by the viewer. (1990: 107; emphasis in original)

Even among viewers with firm political opinions about nuclear energy who were viewing programmes that shared their viewpoint, criticisms were raised when the programme seemed to be 'unbalanced'. Corner, Richardson, and Fenton also noted that viewers differed considerably in terms of whether they perceived it necessary or valuable to employ emotional, persuasive strategies in the film. Viewers differed in their opinion about the value of '*affective* properties of televisual texts—potent visualizations, dramatic simulations, eloquent personal narratives' (1990: 108; emphasis in original). However, the study showed that the affective dimension was powerful for all the groups, even those who were critical of such techniques.

Decoding Fiction

Another study influenced by the encoding–decoding model was Sut Jhally and Justin Lewis's audience research on *The Cosby Show*. The study was funded in part by Bill and Camille Cosby and was widely publicized in the USA. The design called for focused interviews in which a single episode of the show would be shown and discussed by people who were already familiar with the show. The sample consisted of fifty-three small friendship or family groups, roughly divided between white and African-American residents, and between middle-class and working-class groups. All were residents of a small New England city.

The study, published in 1992 as the book *Enlightened Racism*, incudes a content analysis of the episode, of *The Cosby Show* as a series, and of the history of representations of black and of working-class characters on US television. The ambitions of the study go far beyond the encoding–decoding model—which makes the study a useful polemic, but also limits its interest in terms of audience study. The preface sets out the scope of the book:

> This book deals with issues of immense political importance. It addresses two critical aspects of our contemporary culture: how our most ubiquitous cultural form, television, influences the way we think; and how American society thinks about race in the post-Civil Rights era. We chose to study audience reactions to *The Cosby Show* because of its position in relation to these two issues. (Jhally and Lewis 1992: xv)

In guiding the discussions, trained interviewers showed the programme and then solicited feelings about the characters and descriptions of the plot's episode, in which conflict ensues after Claire discovers her son Theo reading a 'girlie' magazine. Interviewers investigated viewers' decodings only superficially, as in determining whether most respondents liked and admired the characters, and whether they found the situations true to life. Instead, questions about the television programme served as a pretext for getting at larger social attitudes. Even when responses did not suggest attitudes towards class and race, these were pursued by the interviewer:

> These innocuous questions often succeeded in opening up the discussion by giving respondents the opportunity to remark on attitudes toward class, race, or gender, attitudes the interviewer could then explore. If respondents were less forthcoming, the interviewer would ask them to comment on these topics—for example, 'How would you feel if the Huxtable family were white?' and 'Would the show be as good if the Huxtables were a blue-collar family?' Because the initial responses to these questions were sometimes ambiguous, guarded, or even misleading, the answers were carefully explored in the ensuing discussion. (Jhally and Lewis 1992: 11)

Jhally and Lewis claim that their initial findings suggested optimism about the ability of whites to accept black characters—even to love and admire them on television.

The interviewees might have suspected the interviewers of false pretences, as questions about *The Cosby Show* turned increasingly to discussions of race relations in the USA. As the interviewers probed attitudes towards race, they received rather depressing answers: the majority of white respondents enjoyed *The Cosby Show* because the characters were neither too black nor too working class, and—even worse—that the show served as a sort of argument against affirmative action. Thus, *Cosby* is implicated in the defence of Reaganomics, the dismantling of affirmative action, and the widening class divide. In the end, they conclude that *Cosby* has an insidious effect on white Americans:

> For many white respondents in our study, the Huxtables' achievement of the American dream leads them to a world where race no longer matters. This attitude enables white viewers to combine an impeccably liberal attitude toward race with a deep-rooted suspicion of black people.

> They are, on the one hand, able to welcome a black family into their homes; they can feel an empathy with them and identify with their problems and experiences. They will, at the same time, distinguish between the Huxtables and most other black people, and their welcome is clearly only extended as far as the Huxtables. (1992: 110)

For black viewers, *Cosby* is pernicious, not because it encourages racist attitudes, as it does with the white viewers, but because it forces African Americans to accept the television industry's position that normalcy means upper-middle-class status, and that a positive portrayal of Blacks necessitates their belonging to a socio-economic strata that the majority of black viewers cannot hope to attain.

Jhally and Lewis's study, then, successfully publicized an argument about the class background of television characters that has been made most notably by George Lipsitz (1990)—that US television has abandoned dramas about working-class characters and, instead, populates its programmes with upper-middle-class professionals, presenting a demographic picture drastically skewed from that of the real population of the USA. *Enlightened Racism* also introduces an important and too often overlooked discussion of the intersection of class and race identities. But the study's empirical findings are troubling on a number of points. Gender is almost completely erased from its major findings. This seems highly problematic in a study dealing with a genre—the domestic comedy—that turns so centrally around gender conflicts, and an episode theme—magazine pornography—that seems to insist on such a reading. While differences between men and women are rarely highlighted in the interviews, stark oppositions between whites and blacks emerge in terms of attitudes. For the white respondents, the interview situation itself seemed a sort of set-up: to say good things about *Cosby*—perhaps even in the spirit of seeming enlightened, liberal—led one down the road to apologism for an entire decade's political failings—including the retrenchment of the welfare state (H. Gray 1992). This may be a cause of one of the problems that was encountered in conducting the *Cosby* interviews: Lewis notes that analysis of the interviews was difficult because of the 'cautious and evasive' responses on the topic of race. The *Cosby* study calls attention to the ways that informants may be cognizant of the interviewers' wish to categorize them in ideological terms, and may wish to resist these efforts. There is also the question of whether domestic viewing of entertainment programmes can be reducible to ideological position. In another book discussing this research, *The Ideological Octopus* (1991), Justin Lewis described the viewers they interviewed to be lacking in critical discourses about television, but this may speak more to the flaws in the adequacy of the decoding model, and to the superficiality of what respondents feel permitted to say in a focus group, as it does to the degree of racism in the United States.

In some ways, the *Cosby* study tells us little about television itself, since the researchers were eager to move beyond the subtleties of audience interpretation to get to the more important, overarching theme of racism. The danger in such a design is that television is used as a mere pretext for conversation and insufficient attention is given to the complexities of television form. Thus the television programme may be reduced to a series of 'messages' (as in the traditional effects paradigms) and themes—aspects of programming that are clearly only a small part of the experience of television viewing and could easily be ignored or rejected by viewers.

On the other hand, as a study of racism, *Enlightened Racism* offers no information about the connection between words and actions, about the different background of the respondents' lives in terms of their interactions with people of different races. Instead, a highly reified picture of whites emerges that bears little feel for the necessarily lived contradictions of race and class relations in the contemporary USA. Indeed, the picture that emerges is one of a dominant ideological discourse holding total sway over television viewers—something quite other than the adoption of Gramsci's theory of hegemony originally proposed by Hall in the encoding–decoding model.

Women Viewing Violence, the last encoding–decoding study I will consider here, offers research designed to avoid some of these simplifications, while relying on a similar format: the single screening, followed by focus group format. Philip Schlesinger, R. Emerson Dobash, Russell P. Dobash, and C. Kay Weaver organized their research to include much lengthier discussions than those in the *Cosby* study. Subjects typically spent seven hours at the screening and discussion; childcare, food, and payment were offered. The topic was violence against women on the screen, and the subjects were shown either an episode of the British soap opera *East Enders*, the US film *The Accused* ('a feature film made for cinematic release but also available to satellite television and video audiences'), the news magazine show *Crimewatch UK* ('a factual programme rooted in realist conventions of crime reporting'), or *Closing Ranks* ('a single play made for television by a recognised documentary auteur') (Schlesinger *et al.* 1992: 23).

The sample was divided both along lines of class and ethnicity and between women who had experienced violent attacks by men and those who had not. The study focused on the interpretation of discourses (especially about gender and about violence) that are both fiction and non-fiction. The researchers performed an extensive quantitative analysis of questionnaire data (from 542 collected questionnaires), and transcribed more than 100 hours of interviews. The authors are reflective throughout about their choice of methods—and, indeed, the nature of the study helps to focus attention on the kinds of suppression that may be common to 'family' interviews, or conventional means of locating informants. For example, they decided to interview women in all-female groups, because the presence of men would be likely to inhibit discussion. They held screenings and discussions in women's centres or universities, since holding discussions in homes, where domestic violence occurs, would be impractical. Finally, the groups were composed of women of broadly similar class and ethnic backgrounds. They add a caution, however:

> Approaches based on consensus or singularity reify groups and place a gloss on group discussions which is often unwarranted. Our view is that typifications and conclusions about readings emerging from group discussions must be based on tendencies or patterns. The outcomes of group discussions are indicative and symptomatic of certain views and lines of argument, rather than definitive. (1992: 17)

Such a combination of defining subjects by identity variables (gender, race, class) as well as by life experiences (personal experience of domestic violence) makes the *Women Viewing Violence* study unusual. The researchers comment that social science and social work research rarely deal in any way with media representations. It could be added that media studies research usually keeps sociological or social welfare issues at arm's length, although some of these are, of course, implicit in thinking about the ideology.

'Viewing televised violence may, for some women, involve the recreation of a painful and dangerous personal experience; for others, it approximates a feared event; and, for others still, it is merely the depiction of a relatively abstract and distant act' (1992: 164). The authors find that their subjects represent 'a patterned diversity'. The most obvious differentiating factor between viewers was the experience—or the lack of experience—of violence. Viewers who had experienced violence 'were more sensitive to televised violence, more subtle and complex in their readings, more concerned about possible effects and more demanding in their expectation of the producers of such content' (1992: 165). 'Considerable finesse' is how they describe the interpretation of television scenes by women who had experienced violence—they were subtle and sophisticated in their readings of the scenes: 'There was also a knowledge of all the excuses offered for male violence towards women and a consequential refusal to play the game of exculpation which contrasted with the greater willingness to do so among some of the women with no experience of violence' (1992: 165).

In contrast to Jhally and Lewis, Schlesinger *et al.* are quite cautious about their findings. Despite the fact that their research produced active, critical, and highly analytical responses from their respondents, the authors pull back at the end from suggesting that changes in routine programming would have a highly significant impact on the lives of women. The study speaks to the question of accountability of media producers, by suggesting how broadcasters might benefit from listening to the women who are quoted in the study, and taking responsibility for the emotional impact of the programming that they deliver. On a different level, they acknowledge that the experience of talking and thinking about television in unconstrained ways may be useful and important. They end, however, with a note of caution:

> Just what kind of impact—for most—can changing how you think about television (or how you consume culture more generally) have on the social, economic and political constraints of your life? The question does need posing, for there is something of a temptation abroad almost to substitute television criticism for political action. We certainly would not accept that the two are identical by any means. Quite simply, there is in such a tendency a temptation altogether to overrate the importance of television. (1992: 173)

The modesty of this conclusion may be due in part to the collaborative nature of the study, in which Dobash and Dobash's expertise on patterns of domestic violence helps to temper the tendency manifested by some media scholars towards overvaluing media effects.

The encoding–decoding model seems to work better for news and non-fiction programmes than it does for entertainment programmes, where it is much more difficult to identify a single message, or even a set of propositions with which audience members could agree or disagree. Drawing conclusions from his *Nationwide* study—and these would apply to fiction programming as well—Morley has astutely recommended 'dropping the assumption that we are principally dealing with the overtly political dimension of communications' and, instead, 'dealing more with the relevance / irrelevance and comprehension / incomprehension dimensions of decoding rather than being directly concerned with the acceptance or rejection of substantive ideological themes or propositions' (Morley 1992: 127). To some extent this agenda is implicit in the move towards studying the contexts of television reception, especially in domestic space. In the following section I turn to three influential studies of media consumption that are based on the study of the domestic sphere rather than specific texts.

Feminist Studies of Domestic Contexts

As feminist scholars have frequently argued, nuclear families are places where gender roles are produced, played out, and challenged. Three of the foundational works in audience studies that take up this argument are Janice Radway's *Reading the Romance* (1984), David Morley's *Family Television* (1986), and Ann Gray's *Video Playtime* (1992). These projects established the significance for media studies of the feminist tenet that the home is perceived as a place of leisure for men and a place of work for women, and media consumption is inextricably linked to gender roles.

In *Reading the Romance*, Radway relied on a key informant, who worked at a bookstore, recommended paperback books to many customers, and published a newsletter discussing the best romances on the market. This key informant put Radway in touch with other women, to whom she gave a survey questionnaire (thus, employing some quantitative analysis in the book) and invited to focus-group interviews.

In some ways, Radway's study borrowed from the 'uses and gratifications' tradition of media research by asking what place these books have in the lives of her informants. Unlike uses and gratifications research, however, the question of textual interpretation—both Radway's own and that of her informants—is at the centre of the study. Radway's book captures many contradictions inherent in the act of women's reading, by sorting out the differing tendencies in the readers' motivations, in the ideology of the texts themselves, and in the use of reading as a strategy to secure leisure time. The prestige of reading as an activity was used by these women to justify the leisure time required to read books and their own release from housekeeping and childcare chores. Radway mounts a subtle argument that women use these deeply conservative books, in which heterosexual romance provides the ultimate meaning to women's lives, to liberate themselves from the conditions of patriarchal marriage. Radway also frames her study of romance novels with an analysis of the publishing industry, and the place of the genre and its women readers in the economics of bookstores and marketing. In this respect, the study exemplifies a strategy that has been only rarely followed (see Shattuc 1997) of combining political economic research with audience research.

In *Family Television*, Morley found that the behaviours of television usage were inextricably linked to family hierarchies and gender roles. Morley had set out to 'produce a more developed conceptual model of viewing behavior in the context of family leisure' (1986: 17). He interviewed eighteen white families in South London: each family consisted of two parents, two or more children, and had a television and VCR in the home. (Informants were located by a market research firm.) Morley argued that television audiences need to be studied in the natural settings in which most media are consumed, and so he elected to study television at home, among family members.

Morley found distinctively different viewing styles reported by men and women, and a great deal of conflict between them concerning the TV. Husbands charged that their wives and daughters talk too much while the TV is on; wives complained that their husbands talk too little. The men Morley interviewed tended to adopt a style of intense, cinema-style viewing; the women were more distracted, tending to do chores at the same time as watching TV—unless they were alone in the house. Morley organized the gender-related themes in the interview material into the following categories: power and control over programme choice; viewing style; planned and unplanned viewing; amounts of viewing; television-related talk; use of video; solo viewing and guilty pleasures; programme type preference (Morley 1986: 146).

Morley found that men watched more television, planned their viewing more, and tended to control what others in the household watched. Women viewed less, often deferred to other family members in the selection of programmes, and enjoyed watching soap operas and melodramas, especially when they were alone. Morley found that his subjects seemed already to like those 'gender genres' designed for them. Women voiced a taste for soaps and movies; men preferred crime shows and sports. Morley's interviews were conducted *en famille*. Using Pierre Bourdieu's notion of cultural capital, he suggests that men were under-reporting their fictional viewing and over-reporting their viewing of news and documentary, based in part on their perception of the differential evaluations of these kinds of television in the system of social distinctions. Women seemed to feel freer to admit that they watched and liked TV.

In *Video Playtime*, Ann Gray reports on interviews with thirty women from predominantly working-class backgrounds who were contacted through a video rental library. Gray focused on a series of interrelated topics: the incorporation of the VCR into the domestic sphere, the gendered division of labour and leisure in terms of use of and attitudes towards the VCR, and preferences for particular genres of video. All the women in Gray's sample had children at home and had husbands who were the household's primary wage-earner.

Gray found that many women reported that the home was a difficult place for them to relax. The women she interviewed rarely took breaks from domestic chores, which lessened their enjoyment of the VCR. Gray explains that the husbands' greater involvement with the VCR was the result of 'a combination of masculine address of VCR advertising, the relative freedom of male leisure time in the home, and male economic power' (1992: 243). For husbands and children, then, the greater freedom from domestic chores made them more likely to watch videos, rent them, plan ahead for recording and time-shifting, and become adept at operating the machine.

Some of the women expressed resentment towards television and video as a deterrent to engaging in more appealing forms of leisure activities, such as going out, and as a barrier to family communication and intimacy. In an interesting research design borrowed from Cynthia Cockburn, Gray asked her subjects to identify various parts of the VCR controls as 'pink' or 'blue', based on whether they would be more likely to be used by male or female members of the family. The timer switch was always blue, with women depending on their husbands and children to operate it, and the remote control device tended to fall into the hands of the male partner or male child. Only the play, rewind, and record functions were 'lilac', being used equally by male and female family members. Gray insisted that the issue of competence in operating a VCR is inextricably linked to domestic labour.

Gray found more differences among women in her sample in the area of preferences for videos to rent and television material. She divided her sample into two groups: 'Early School Leavers' and 'Later School Leavers and Graduates'. She found starker gender differences in the second group, who tended to evaluate negatively a taste for romance, melodrama, and trash TV. Following Morley, Gray found that her subjects correlated feminine tastes with adjectives such as 'soft, soppy, fantasy, silly, fictional'. They correlated masculine tastes with such words as 'hard, tough, real, serious, and factual'. The more highly educated women in her group tended to share preferences with their partners: 'These women claimed to have similar preferences in programmes and films to their partners and the majority of them were keen to distance themselves from soap opera, particularly American products . . . and to align themselves with "quality" products' (1992: 160). Gray suggests that Bourdieu's work—which I review in the next section—can be used to explain the tendency for these educated women to have achieved 'an aesthetic disposition', which included a distanced objectivity with regard to television products, and, in particular, a distaste for soap operas, and a concern for their ill effects.

Ien Ang and Joke Hermes have criticized these studies for offering essentialist renderings of gender. They have argued that 'an individual's gendered subjectivity is constantly in the process of reproduction and transformation. Gender does not simply predetermine media consumption and use' (Ang 1996: 116). The portraits of domestic life constructed by Morley, Radway, and Gray seem very much at odds with the kind of formulation Ang postulates when she writes: 'Gender identity, in short, is both multiple and partial, ambiguous and incoherent, permanently in process of being articulated, disarticulated and rearticulated' (1996: 125). From this perspective, the families in *Family Television* represent a feminist nightmare: father holds the remote control and imposes his programmes on everyone else; mother desperately, guiltily, sneaks time to watch a weepie. There is a conflict, then, within media studies between theoretical frameworks, modes of doing research, and methods. Radway, Morley, and Gray's works construct their notions of subjectivity through empirical studies of media use by adults with children living in nuclear families—a group that is perhaps more fully engaged in the social reproduction of gender on a daily basis than are single adults. Their research is based on reports of behaviours, interviews, and discussions of everyday practices (a kind of data that is likely to reflect many commonsense notions of gender difference)—a subject defined through a materialist approach to audience research. Ang and Hermes construct media subjectivity through postmodern *theories* of ethnography, through theoretical discussions informed by Michel Foucault, Jacques Lacan, Judith Butler, and others—based at times in textual analysis, but rarely in empirical field research—and argue that gender is constructed as contradictory and shifting.

The problem of gender essentialism in these studies is as much methodological as it is theoretical: more intensive ethnographic methods might have uncovered more contradictory and shifting positions among Morley, Gray, and Radway's informants; Radway and Morley have advocated such methodologies in subsequent work. To some extent, the question of whether the discussion of gender in these studies was mechanical and essentialist is an empirical one. More research needs to be carried out in the field to note the degree to which traditional gendered roles in the family are flexible, shifting, and multiple. In the chapters that follow I sketch research procedures that contribute a more complex view of gender, media consumption, and domestic contexts by using longer contact periods, implicating institutions as well as individuals in their analysis, and developing the analysis of domestic space as a context.

Bourdieu on Television

Pierre Bourdieu's work, with its emphasis on the differential distribution of cultural tastes and on the embeddedness of tastes in the habitus, the material—and domestic—structures of everyday life, has exerted a major influence on television studies since the publication of *Distinction* in English in 1984. Bourdieu's empirical research and his theories of the role of aesthetic distinctions in the construction of social hierarchies have resonated with questions about television audiences, the importance of the domestic sphere as a site for the inculcation of tastes and a place of aesthetic consumption, and the accentuated awareness of the variability of interpretations of aesthetic texts. *Distinction* has helped scholars conceptualize television in relation to other cultural forms, and to force them to think about the relationship between tastes for particular kinds of television genre and class position. Bourdieu's best-known argument is that 'art and cultural consumption are predisposed, consciously and deliberately or not to fulfill a social function of legitimating social differences' (1984: 7).

In *Distinction* and other work, Bourdieu has had little of a theoretical nature to say about television (more recently he has published a diatribe about the impoverished nature of television programming as compared to academic discourse, and its detrimental impact on democracy: see Eakin 1997). Yet Bourdieu has proven useful to television scholars because he pays particular attention to forms of culture stigmatized by intellectuals and the bourgeoisie: forms considered vulgar rather than refined, emotional rather than mental / intellectual, expressive rather than aesthetically distanced. Of course, such concerns match rather well with most popular television genres, especially sensationalistic forms: talk shows, soap operas, sports, wrestling, crime dramas, music videos, etc. Bourdieu has helped to bring to the foreground ideas about the ambiguous social status of television as a cultural form that is at once widely accessible and widely deprecated.

Bourdieu focused attention on the role of education and the influence of 'cultural capital' on taste, the selection and valorization of certain cultural forms. In his introduction, he formulates this in the widely quoted statement 'Taste classifies, and it classifies the classifier' (1984: 7). These distinctions are used to legitimate the privileges of those with more education and more money, who envision themselves as superior to those whose tastes differ from their own. Bourdieu emphasizes that these distinctions are just as present in the selection of novels to read or pictures to hang on the wall of one's home as they are in choices of food or hairstyle. His account focuses on the relationship among types of goods, and argues that the meaning of any given commodity (such as a television programme) derives from its similarities to and differences from other commodities in society (live performances of opera or ballet; football games). Increasingly, society requires consumers to understand and manipulate complex meanings and connotations attached to consumer goods and commodified cultural forms, so that they may choose to make the right impressions—and so that they may avoid mistakes. This can involve complex negotiations in the linking of cultural forms to social status. Emulation involves a double movement: imitation of those richer, and differentiation from those poorer or less 'refined.'

To a large extent, the attention to strategies of decoding texts in audience studies has been an excavation of forms of cultural capital unrecognized by prior forms of communications research—thus contributing to the sense of television as a complex and even fascinating media form with its own codes. In one of the earliest examples, Charlotte Brunsdon pointed out that:

> Just as a Godard film requires the possession of certain forms of cultural capital on the part of its audience for it to 'make sense'—an extra textual familiarity with certain artistic linguistic, political and cinematic discourses—so too does . . . soap opera . . . the narrative strategies and concerns . . . call on the traditionally feminine competencies associated with the responsibility for 'managing' the sphere of personal life. (Brunsdon 1983: 80)

This kind of argument led to observations by Robert Allen, Dorothy Hobson, Tania Modleski, Andrea Press, and Seiter *et al.*, that members of the audience who despised soap opera, often were simply lacking in the cultural capital required to read the text adequately—and that the low status of soap opera audiences could best be explained as a result of a social structure which routinely placed working-class and feminine forms at the bottom.

The concept of cultural capital has been widely appropriated throughout television studies. Bourdieu has described his project 'to grasp capital . . . in all of its different forms, and to uncover the laws that regulate their conversion from one into another' (Bourdieu and Wacquant 1992: 118). Four different forms of capital were identified by Bourdieu: economic, cultural, social, and symbolic. Economic capital includes financial resources of various kinds, and encompasses the bases for most traditional definitions of class, such as by income level. Cultural capital adds to this an embodied state of tastes, preferences, and knowledge, ranging from educational credentials, to preferences in music, to embodiments of femininity. Social capital consists of networks, connections, group memberships, familial relationships: 'Social capital is the sum of the resources, actual or virtual, that accrue to an individual or a group by virtue of possessing a durable network of more or less institutionalized relationship of mutual acquaintance and recognition' (Bourdieu and Wacquant 1992: 119). Finally, symbolic capital is the form achieved when the economic, cultural, and social capital are recognized as legitimate and institutionalized.

Individuals could accumulate large stores of cultural capital in relation to television—for example, knowledge of twenty-five years of a soap opera's history, the names of actors, gossip about the stars, or by watching the evening news every night for decades. This cultural capital could be used in the currency of friendship or polite conversation with neighbours or family members, but unless it can be converted to symbolic capital (to get a job as a TV studies professor, or to land a job at a television network), it has not been translated or exchanged for symbolic capital.

It would be fair to summarize the influence of Bourdieu on television studies by saying that there has been considerable attention to cultural capital, and some attention to social capital in ethnographic audience studies tracing TV as a part of social relationships, but very little attention to, or indeed cognizance of, symbolic capital in relation to these. As Beverley Skeggs explains:

> Symbolic capital is powerful capital: it brings power with it. If one's cultural capital is delegitimated then it cannot be traded as an asset; it cannot be capitalized upon (although it may retain significance and meaning to the individual) and its power is limited . . . Most representations of working-class people contribute to devaluing and delegitimating their already meagre capitals, putting further blocks on tradability, denying any conversion into symbolic capital. (1997: 11)

The celebration of forms of cultural capital involved in the appreciation of television programmes, a tendency perhaps best exemplified by the work of John Fiske and Henry Jenkins, is a position that Bourdieu himself explicitly opposed. Bourdieu is adamant about the necessity of a means of translating cultural capital into social capital for any material rewards to accrue to an individual. Clearly, only a small group of professionals (academics and media producers, usually backed by forms of middle-class education and credentials) can 'convert' the cultural capital of knowledge about television into social capital. Even when researchers identify themselves as fans (Hobson, Jenkins, Bacon-Smith), there is a considerable difference in cultural capital between interviewers and informants. Brunsdon has pointed out that such work has avoided confronting questions about the quality of the media itself:

> Only the inheritors of legitimate culture, researching other people's pleasures, pleasures they may well share, can afford to keep quiet about the good and bad of television. They—we—through years of training have access to a very wide range of cultural production. Watching television *and* reading books about postmodernism is different from watching television and reading tabloid newspapers, even if everybody concerned watched the same television. (Brunsdon 1990: 69)

Brunsdon argues that too often, by validating the pleasures of television viewing, scholars fail to make any demands for different, even better forms the audience might want.

Here, Brunsdon is in keeping with the spirit of Bourdieu's work, which finds any celebration of the popular bankrupt and warns that: 'To act as if one had only to reject in discourse the dichotomy of high culture and popular culture that exists in reality to make it vanish is to believe in magic . . . What must be changed are the conditions that make this hierarchy exist, both in reality and in minds' (Bourdieu and Wacquant 1992: 84). Like Brunsdon, Bourdieu has insisted that the intellectual's peculiar place in the system of social distinctions—a set of predispositions that make him a poor (or, rather, an inevitably interested) party in discourses about television. Thus, Bourdieu has perhaps inspired one of the recurring critiques of some audience research of the 1970s and 1980s. For example, Jostein Gripsrud comments that, too often, audience researchers have adopted a 'pro-television' position and failed to recognize their position as 'double access' audiences: 'Intellectuals now have access to both high and low culture, they are "double access" audiences; the majority of "ordinary people" have only access to "low" or popular culture. Our double access is a class privilege, a benefit of education, from which we cannot escape' (1995: 125). Audiences who have been denied access to more elite forms of culture tend to rely heavily on television as a media form, but are vulnerable to recognizing a judgement from above of television as trash, as a waste of time. Forms of apology or guilt about television viewing by heavy viewers can be viewed as a symptom of 'symbolic violence', which Bourdieu defines as 'the *violence which is exercised upon a social agent with his or her complicity*' (Bourdieu and Wacquant 1992: 167; emphasis in original). Bourdieu sees the failure to recognize symbolic violence as a chronic failing of sociological researchers, a blindspot in the goal of reflexive sociology: 'Intellectuals are often among those in the least favorable position to discover or to become aware of symbolic violence, especially that wielded by the school system, given that they have been subjected to it more intensively than the average person and that they continue to contribute to its exercise' (Bourdieu and Wacquant 1992: 170). Critics of Bourdieu have suggested that his model is functionalist, overly pessimistic and deterministic, and puts forward the dominant ideology as an all-powerful and universally accepted standard. John Hall explains one aspect of this critique when he points out that Bourdieu assumes that everyone recognizes the legitimacy of distinctions handed down 'from above': 'To describe any one social group's calculus as the effective one is to confer legitimacy to a calculus that, as Bourdieu recognizes, remains in play with others . . .' (Hall 1992: 279).

Another important limitation of Bourdieu's work involves the ways in which it universalizes the quite specific social hierarchy of French society. Clearly, much adaptation is needed of Bourdieu's model to British and probably still more to US culture. Sociologists Michèle Lamont and Annette Lareau have predicted that American legitimate culture is 'less related to knowledge of the Western humanist culture, is more technically oriented (with an emphasis on scientific or computer information) and more materialistic' (1988: 66). They suggest that purchasable—rather than culturally acquired—signals of legitimate culture may be more acceptable and may be granted more weight—in Bourdieu's terms, more easily converted to symbolic capital—in the United States than in France.

Bourdieu's greatest value to television audience research may be his enthusiasm about the practice of empirical sociology, his urgent calls that such field research must complement theoretical work on the sociology of culture. Thus, the move outwards to the audience is in line with his notions of reflexive sociology, a form of research which practises, at best, 'the inclusion of a theory of intellectual practice as an integral component and necessary condition of a critical theory of society . . .' (Bourdieu and Wacquant 1992: 173).

Speaking Subjects

Bourdieu's research has often been based on large-scale survey research. It has been open to the same criticism as that lodged at traditional media effects research, in that its use of research procedures do not do justice to the increasing theoretical sophistication about language that informs so much Marxist, postmodernist, and poststructuralist theory. Speech is the primary form through which researchers gain access to information about media audiences. This fact links audience studies with theories about language and subjectivity that have brought about dramatic reappraisals in the humanities and social sciences.

Much of the ethnographic research on audiences is grounded in theories of subjectivity, based in part on the work of Louis Althusser and Jacques Lacan. In this theoretical tradition, subjectivity is a term for consciousness, and 'subject' replaces terms such as individual, person, citizen. The term subject connotes a certain degree of passivity, implying one who is subjected to something (namely ideology and unconscious processes) rather than a free individual acting upon the world (a conception more in line with uses and gratifications research). Some audience researchers have stressed the multiplicity of subjectivities. This position implies a political understanding of differently defined and created identities, such as gender, class, race, ethnicity, sexual orientation; and of the unconscious as well as the conscious mind. The stress here is on the intentional and unintentional nature of the subject's speech and way of making sense of the world. Language cannot be treated as a perfect match with the intentions of the speaker, nor as a realist system of representation, a transparent, immediately comprehensible vehicle for communication.

In the poststructuralist view, language is theorized as a kind of prison. Language is not a free, open form that expresses us perfectly; rather, it preexists us as individuals, and all our utterances are trapped within structured, conventional, ideological systems of language—or discourse. In its current usage, the term discourse carries the implication for speech governed by social, material, and historical forces, which disallow certain things from being said or even thought, while forcing us to say certain other things. Many scholars use it in Foucault's sense to refer to a set of complex, multilayered texts that determine and limit what can be said or known about certain subjects, and therefore serve particular interests in the power structure of society. Discourse is not 'free speech'. It is not a perfect expression of the speaker's intentions. Indeed, we cannot think of communicative intentions as predating the constraints of language at all.

The methodological implications of this theoretical work are that what people say when talking about the media cannot be taken at face value. We cannot assume that what subjects say in an interview reflects individual, idiosyncratic views, or that what is spoken is all there is to be said on the subject. First, our subjects may not have access to all that might be going on with their media consumption, because of the role of the unconscious. Second, media tastes do not simply reflect identity, but are actually constitutive of it. Therefore, one of the things we would expect to hear from subjects is the reiteration of certain prior existing discourses on the self, society, politics, and gender.

A somewhat different perspective on the problems of language in qualitative research is offered by ethnomethodologists and conversation analysts, who focus on the ways in which speech events construct and affirm reality for speakers. Their work suggests the importance of looking at the conventions of speech and the commonplace understandings of what is happening that are operative in interview situations. Ethnomethodology would call attention to how researcher and subject 'do interviews' and at the tacit procedures that rule the situation. Ethnomethodology focuses on such speech situations and practices as 'contingent ongoing accomplishments of organized artful practices of everyday life' (Garfinkel 1976: 11). According to Lindlof, conversation analysis is a branch of ethnomethodology that focuses on 'such features of ordinary talk as the way conversations open, the order in which speaking turns occur, the sequencing of utterances, the repairing of problems, reflexive expressions about the talk, and in general, the manner in which spontaneous conversation displays the appearance of a polished performance' (1995: 39). This phenomenological work has so far been less influential on the development of media audience research than have theories of discourse and subjectivity, but it deserves serious consideration by audience researchers.

David Buckingham has usefully summarized the issues as they relate to talk about television concerns:

> Clearly, individual users of language have no option but to select from among the available linguistic resources, which are already structured in particular ways. Language therefore cannot be seen as merely a neutral vehicle for 'attitudes' or 'beliefs', or a product of mental entities or processes. At the same time, subjects use language to construct versions of social reality: to a large extent, what people talk about is constructed in the process of talk itself. These versions of reality are consequential, in the sense that they perform specific social functions or purposes. In these respects, then, language is both constructed and constructive. (1991: 229–30)

At its worst, audience research simply ignored these theoretical developments and fell back on a realist treatment of language, analysing transcription of speech as a pure and direct expression of the mind of the subject. At its best, this research accepts that 'the audience' is ultimately unknowable in some totalizing way, yet strives for a research design that maximizes an awareness of the researchers' own role in moulding what is said and how it gets said. There is considerable difference of opinion as to how much contact is necessary before a group of informants feels comfortable enough to act and speak naturally in the presence of a researcher. Lull (1988), who takes a realist view of language and whose work fits into the US uses and gratifications paradigm, reports that families quickly felt at ease and behaved normally, despite the observer's presence. Others—and I would place myself in this camp—consider the impact of the researcher's presence to be a continuing and strongly influential factor in shaping the interaction, and limiting what is said.

Psychologist Valerie Walkerdine's work, which has been strongly influenced by Lacan and Foucault, may be the best example of field research that focuses on the unspoken, on the gaps between desires and their expression. Walkerdine has frequently noted the ways that middle-class families are more talkative about the media than working-class families, where practices of 'talking things out' and therapeutic conversations are less common: 'The target of regulation, then, is especially the working-class family . . . who do not critically discuss programmes and use television to foster family harmony or as a means of avoidance rather than a tool for advancement by beginning "viewing TV more critically"' (Walkerdine 1993: 75–6). Holding a theoretical understanding of the limitations of making subjects speak is important; but it is also true that media researchers, if given more time and repeated contact with subjects, could employ some practical strategies to cope with silences and to recognize the difference between a strategic refusal to speak and a poorly facilitated interview. It is important to give subjects more than one occasion for communicating with the researcher and more than one means of doing so. For example, women are often silent in a group interview, but more loquacious in a one-to-one situation. Some respondents may not wish to say things out loud that they would be willing to put in writing. When the interviewer is not using the informant's first language, the informant may feel more comfortable speaking through an interpreter than venturing to speak up in a second language.

Most work on audiences has relied exclusively on verbal transcripts, usually quoted without description of the context in which a statement was made. Dependence on a printed transcript results in a tremendous loss of information in terms of the non-verbal communications that accompany speech, such as eye movements, facial expressions, hand gestures, and head and body movements, as well as tone of voice, rate of speech, loudness or softness, etc. One of the best exceptions to this is the book *Children and Television* by Bob Hodge and David Tripp (1986). The authors collected data by videotaping and audiotaping children while they discussed particular cartoons. Careful attention was paid to non-verbal communication. This is an especially rich method of study for children's language, where vocabulary may be limited but communications are very rich in terms of non-verbal and phatic elements. It is also especially important to look at the power differential between the adult interviewer, who has authority over the children, and the children.

Hodge and Tripp's transcript of the group interview tells one story; the videotape tells another. Hodge and Tripp's discussion benefits from a very finely tuned sense of how the social situation produces what can be said about the cartoon being studied. Similarly, in David Buckingham's interviews with children, he found that:

> they perceive the [interview] context as one in which a relatively 'critical' response is at least appropriate, and possibly even required. The 'critical' discourse serves a dual purpose: it enables the children to present themselves as 'adult', for the benefit of each other and myself; and it provides a means of refuting what they might suspect adults (including me) to believe about the influence of television upon them. (1993: 231)

Hodge and Tripp pay careful attention to how the children's inflections, their use of rising and falling tones, and 'babyish' voices to communicate non-seriousness, significantly shape the meaning of their speech. Throughout the group interviews, boys and girls interact with each other, and individual children become leaders within the groups. In analysing videotapes and transcripts of these discussions, it became apparent that in many instances boys silenced girls, adults silenced children, and interviewers silenced subjects—through non-verbal censure of some remarks (glances, laughs, grimaces), by wording questions and responses in certain ways, or by failures to comprehend each other's terms.

Hodge and Tripp's analysis represents a type of methodology that is very rich in complexity, but—unfortunately—very time-consuming, requiring as it does a careful study of the videotape and notation of all of the children's actions. All kinds of prohibitions exist on the ways that children will discuss television with adults because of status differences, and children's knowledge about adult disapproval of popular culture. Hodge and Tripp accept that researchers simply will not discover everything that children think about TV, but careful attention to non-verbal as well as verbal clues gives many clues to the ways that the contexts of research produce certain forms of speech from children. Children present a particularly obvious case of the importance of non-verbal communications, but researchers dealing with adults need to be similarly aware of the powerful communicative role of the non-verbal in interview and conversational situations.

Ethnography's Other

The origin of ethnography is rooted in colonialism. Historically, ethnographies have been written by Europeans and Americans documenting their experiences among people living in Africa, Asia, or Native American cultures. James Clifford and others have mounted a critical challenge to traditional ethnography's implicit insistence on scholarly experience as an unproblematic source and ultimate guarantee of knowledge about a specific culture or cultural process. Clifford has rejected 'colonial representations' as 'discourses that portray the cultural reality of other peoples without placing their own reality into jeopardy' (1983: 128).

Do audience studies also construct a sort of colonial representation? Is there an 'Other' who is the subject of audience ethnography? Most audience research exists in an ambiguous relationship of alterity to the culture: when researchers investigate media use are they venturing into 'foreign' lands or not? Valerie Walkerdine has stressed the ways that fantasies of the 'Other' play a major, if usually unwritten role in social science research, and that, for mass media research, class has been one of the most important structuring differences between researcher and researched. Walkerdine's case studies of families analyse the interplay between video watching and casual conversation. She includes, to a much greater degree than other researchers, or than would be acceptable according to the conventions of most social scientific writing, autobiographical material as well as an analysis of the way her own family background compels her to 'read' the family interactions in certain ways. Walkerdine seeks to avoid 'exoticizing' the Other through this approach. She is harshly critical of the representation of the audience in most social science:

> The audience for popular entertainment, for example, is often presented as sick (voyeuristic, scopophilic) or as trapped within a given subjectivity (whether defined by the social categories of class, race and gender or by a universalized oedipal scenario). What is disavowed in such approaches is the complex relation of 'intellectuals' to 'the masses': 'our' project of analyzing 'them' is itself one of the regulative practices which produce *our* subjectivity as well as theirs. We are each Other's Other—but not on equal terms. Our fantasy investment often seems to consist in believing that we can 'make them see' or that we can see or speak *for* them. If we do assume that, then we continue to dismiss fantasy and the Imaginary as snares and delusions. We fail to acknowledge how the insistent demand to see through ideology colludes in the process of intellectualizing bodily and other pleasures. (Walkerdine 1990: 199–200)

Many studies have focused on women from working-class or middle-class backgrounds (Radway 1984; Morley 1986; A. Gray 1987; Seiter *et al.* 1989; Press 1991). In this work the projection onto the audience as an Other primarily involves class issues, although there is often a strong component of identification and even solidarity between the feminist researcher and her informants.

Most US and UK audience research has involved white researchers and white informants, although the researcher frequently apologizes for the lack of diversity in the sample. As Jacqueline Bobo and I have argued elsewhere, the homogeneity of the samples does not occur accidentally: white researchers have not been alert to the self-selection at work here (Bobo and Seiter 1991: 290–2). People of colour may not have sufficient trust in or comfort with white researchers to participate in audience research. The tendency for qualitative interviews to be carried out in people's homes may also dissuade some from participation. The strain of caring for children and working long hours, sometimes at multiple jobs, will mean that many more impoverished informants will simply not have the time to be interviewed.

Bobo's study of the reception of the film version of *The Color Purple* among African-American women is an important exception to the all-white sample. In Bobo's focus groups, African-American women reported how much they valued being interviewed by an African-American researcher. In a follow-up interview, Bobo mentions to the group that her research has been criticized as representing an atypical sample because her subjects were so articulate:

> Once again, the women displayed a shrewdness about their status in society and about the way black women are viewed by others. One of the women wondered if the critics knew that a person could be intelligent but not necessarily well educated . . . Still another commented that many people react with surprise when they hear black people speak sensibly because, too often, that is not what is allowed to be presented in a public forum. She then asked a rhetorical question toward which the other women responded with spirited agreement: 'Don't you think we would have come off sounding stupid if someone other than another black person was doing research on us?'. (1995: 132)

Bobo's discussion is an important example of the way that media audiences may be critical of the ways they are characterized by researchers. Bobo's research design, returning to the focus group to discuss her 'findings' and their reception by an academic audience, allowed her to include such criticisms in her discussion. One of the continuing struggles within audience research is to expand the diversity of the researchers and the informants and to heighten self-reflexivity about the impact of racism on our knowledge production; the perception of various methods by groups; and the politics of exploitation of research subjects by researchers. As it stands, we know much more about white, middle-class audiences in the UK and the USA than about any other groups. Media audience researchers are academics with specific social, class, and cultural backgrounds, who frequently leave their normal places of work and residence to seek out 'the field' and learn about groups with social and cultural backgrounds different from their own. These differences are of a lesser magnitude than those between first world / third world ethnographers, but they are present none the less. In the three chapters that follow, I present case studies that explore these relationships between interviewer and subject, and a variety of approaches to the problem of self-reflexivity in audience research.

3

Feminist Methods: The Parents' Support Group

You know, I think you hit points in your life where you go, I do anyway, where I'll watch The Tonight Show, *you know, and be up until 12.30, and then I'll start watching* Letterman *and I enjoy it, because it's like, it's almost a form a rebellion, where, I don't have any leisure time, so I'm just gonna take it, you know! And if it means it's going to make you really tired—to me, it's worth it. It's worth the trade-off. Because I feel like sitting there watching* Letterman, *and I still get his jokes, and I feel like 'I'm not dead!' I still understand adult things, I can flow with that subtlety. I just go through phases like that.*

Tess, employed mother of 5-year-old boy

FROM 1988 TO 1993, I belonged to a parents' support group organized by a private social service agency whose mission is to support first-time parents and prevent child abuse. A facilitator contacted all new parents while they were in the obstetrics ward at the hospital to invite them to join a monthly meeting. The sixteen parents (fifteen white and one Asian-American) who comprised the group lived in neighbourhoods across the city, worked in different kinds of jobs, and were introduced to each other for the sole reason that our children happen to have been born within the same three-month period. There was, of course, a considerable self-selection factor at work in joining such a group: these adults were committed to and sufficiently worried about their children to pay a small membership fee (twenty-five dollars), attend monthly meetings, and host some of the meetings at their house or apartment (where drinks and food were usually pot luck).

The group had met about once a month for two years when I asked them in 1990 if they would be willing to become a focus group for my research. The reaction of the group was positive, enthusiastic. I conducted three taped group interviews over a three-year period, while taking notes on discussions about television that naturally occurred each month during our meetings. In our group discussions (for which I ordered pizza and paid two teenage girls to supervise the children), children's television viewing fitted easily into an agenda of perceived problems facing parents, taking its place alongside feeding, toilet training, and fire safety. The very ease with which television was accepted as a suitable topic was striking. It was deemed by everyone to be eminently appropriate—and was clearly already cast as a 'problem' warranting discussion, and, if possible, expert advice. Whenever we talked about it, there was an outpouring of often anguished feelings about its impact on child-rearing and marriage—a type of discourse at odds with the optimism of television scholars such as John Fiske, Henry Jenkins, and Lisa Lewis about popular pleasures and television viewing as resistance.

The greatest benefit of this lengthier contact period—and its advantage over research designs such as Ann Gray's and David Morley's—is the much deeper familiarity with the families' everyday lives that I developed over time. This type of research design helps to identify the changing material conditions of women's lives—to offer a series of snapshots over time of television viewing as embedded in processes of housework and childcare which are themselves in flux, a subject of struggle. The impressions gathered through repeated contact allowed me to present a less reified version of television viewing and family life—although in important ways my work elaborates on the ways that mothers and fathers experience gender in a rather fixed and unidimensional way.

Such research allowed me to connect the ethnomethodological analysis of talk about television with the wider ideological discourses of child-rearing, of marriage, of cultural distinction, which I have so often observed saturating women's conversations. Knowing these women better, it was possible to begin to correlate their receptiveness to advice literature, for example, with their material circumstances of employment, childcare, domestic space, and the construction of time. A longitudinal and more extensive contact period is the best way to connect up attitudes towards television with a meaningful composite of information about the family's social position.

As a participant in the group, which had already been formed on the basis of interests we shared apart from my research, I had an opportunity for both formal and informal exchanges with other members, to gather solicited and unsolicited comments on the topic of television. That process allowed me to capture the ways that the group negotiated a particular kind of discourse on television, and to comment self-reflexively on the methodology I employed and my commonalities with the other women in the group as a mother, as well as the differences that derived from my status as an intellectual with a relatively well-paid, comfortable job.

My research methods owe much to discussions among feminist sociologists and anthropologists about what constitutes feminist sociological enquiry (Smith 1987, 1990a; Harding 1987). A resistance to positivism and a consciousness of the ways social science has objectified women have motivated the exploration and examination of the relationship of researcher and subject. In much influential feminist work the motivation is clearly to mount an argument that brings to light various aspects of women's oppression (Harding 1987). This characteristic of feminist research on an epistemological level has also meant that less conventional research procedures—such as small sample sizes and close, 'biased' relationships between researcher and subjects—have gained greater acceptance than in conventional sociology.

Research methods identified as feminist have frequently involved an identification between researcher and subject (construed as a relationship of solidarity, of sympathy and respect). In media studies, one way that such solidarity was expressed was through the investigation and to some extent valorization of despised women's genres such as the soap opera, in such research as Dorothy Hobson's sympathetic interviews with housewives about soap opera viewing (1982); Andrea Press's discussions of television with working- and middle-class women (1991), and, more recently, with pro-life and pro-choice groups (Press and Cole 1995, 1998); Mary Ellen Brown's work on soap operas and gossip (1990); and Joke Hermes's conversations with women about their consumption of women's magazines (1995).

Elsewhere, Gabriele Kreutzner, Eva Warth and I have written about the ways in which, in the soap opera project, status differences among women were often lost in our group interviews that were part of this same project: we were a kind of fantasized sisterhood—just girls getting together talking about TV (Seiter *et al.* 1989). In those interviews, we found that it was in fact easier when interviewing women, as opposed to men, to steer them away from extraneous topics and toward others, especially those that echoed concerns of feminist theory; we could then prove the nascent feminist reading tactics—the semiotic guerrilla warfare—performed by these ordinary fan / heroines. It was on the basis of such research, in part, that feminist interest transformed soap opera into a very productive field for academic enquiry and into an instance of empirical audience research influencing theories about the ideological, social, and psychological mechanisms of audience response (Brunsdon 1993).

The closeness of feminist researcher and subject has in some instances manifested itself in methods reflecting a commitment to sharing the researcher's perspective with the subjects (Acker, Barry, and Esseveld 1991; Acker 1973). In the parents' support group, I was able to share, debate and discuss my work with them (Seiter 1993). At its most intensive, this form of feminist research construes the project as a collaboration or a research partnership between researcher and subject. This trend has been more common in sociology than in media studies, but it was used extensively by folklorist Camille Bacon-Smith in her ten-year relationship with *Star Trek* fans and her formation of close personal ties with that group, which included developing friendships, meeting on vacations together, and reading each other's fan fiction.

To some extent, there has been a re-evaluation of the kind of sympathy and solidarity that characterized feminist research methods in the 1970s. This has expressed itself as an increasing self-reflexivity, but also a scepticism about the researcher's ability to overcome differences of race and class based on gender solidarity, and even a suspicion of the good feelings that often characterize interviews by feminists with their subjects. As Sherryl Kleinman and Martha A. Coop usefully remind us, emotions are rarely scrutinized when they are positive, and they are usually written out of ethnographic accounts based on the normative ideal that emotions contaminate scientific objectivity. Good feelings, in other words, are a significant influence on the interaction of interviewer and subject. Kleinman and Coop are right to argue that fieldworkers may perhaps need to be most cautious when feeling empathetic, friendly, and close to their subjects. 'When relations are smooth and we think we have achieved the right amount and kind of empathic feelings, we need to be the most alert about the analytical import of our feelings. Because we have been taught that sympathy sentiment is a prerequisite for fieldwork, we are unlikely to recognize our good feelings as data' (1993: 46).

They suggest closely analysing questions about the subjects' motivations for agreeing to be studied, their investments in gathering sympathy from the interviewer, or gaining legitimacy for their activities or opinions (1993: 47).

All feminist researchers need to be wary of what Judith Stacey has called 'the delusion of alliance' (1988: 22). Women in group interviews produce such an alliance when they minimize rather than accentuate status differences in conversation, with the intention of making things go smoothly and everyone appear friendly and supportive. Feminist television scholars have shown an increasing awareness of the limitations of this kind of interviewing. Ann Gray suggests that the researcher be sceptical of situations based on shared media experiences, especially fan involvement, on the basis that the female researcher 'has access to quite powerful institutions and intellectual capital'. Nevertheless, Gray concluded that her shared position of class origin with her subjects was 'quite crucial to the quality of the conversations I had with the women and that the talk that ensued was, in most instances, enriched by that shared knowledge. To put it quite directly, I am a woman in my study' (1992: 34). Gray's position is a tricky one, however, because it leads her to insist that she could only include in her sample white women of the same ethnic background. Surely, researchers must be willing to engage subjects whose experiences are different as well as the same as their own, and to interrogate and even pursue situations in which some discomfort exists. Tensions based in inequalities between various ethnic and racial groups will be crucial to consider, especially since the majority of academics in the fields of media studies and cultural studies are white. Seeking subjects who live outside of nuclear families and couple relationships, increasing the number of lesbian subjects, of women of colour, and women under twenty and over fifty years of age would help avoid the tendency to reify the nuclear family in audience research—a topic I will return to at the conclusion of this chapter.

My research with the support group is perhaps most similar to that of feminist psychologist Wendy Hollway, who interviewed friends about heterosexual relationships and romantic love, at times adopting the role of 'informal counsellor' to the group. Hollway has described doubts similar to those I experienced about this sort of design:

> [T]his was because enjoying myself talking to people who wanted to talk to me did not feel like data gathering. It is only now that I can look at it quite the other way round and say that I succeeded in forging a valuable method: that is, to talk with people in such a manner that they felt able to explore material about themselves and their relationships, past and present, in a searching and insightful way. I did not feel skillful because it came so easily. It was easy because the research participants were people like me and we were continuing an activity that was a vibrant part of our subculture at the time. Now I can believe that this made for good research practice. At the time I was anxious that it was a bit of a con. (Hollway 1989: 11)

Hollway has offered a defence of this kind of method based on the poststructuralist perspective that what people say about themselves in an interview is already constructed by a variety of discourses, and that if merely descriptive interviewing is used, there is a tendency to accept individuals' accounts of their experience at face value, to reify the moment of the interview, and to fail to probe the many contradictions inherent in women's everyday lives and how they explain them. Doing an extended research project with people with whom one is well acquainted, on the other hand, allows for the challenging of these accounts:

> Consciousness is not an unmediated product of experience, because meaning intervenes, and meaning is not neutral. It has a history within power relations. When someone gives an account of her experience, some meanings are more anxiety-producing or ego-threatening than others, and through defence mechanisms, they can be avoided. An analysis of accounts that does not acknowledge this—I am tempted to say that *avoids* it—can only reproduce knowledge that is a product of those repressions. (Hollway 1989: 45–6)

In much, but by no means all, feminist research there has been a high degree of self-reflexivity about the kinds of meaning attached to the research or interview situation, and the ways that other kinds of status are brought to bear on that relationship. Sandra Harding rightly argues against the definition of a distinctive feminist research method, since she defines method as simply a set of research procedures, evidence-gathering techniques, and sees feminist research as owing more to methodology (a theory and analysis of how research should proceed) and epistemology. For Harding, a requirement more important than the employment of particular methods, or privileging of particular research questions, is the insistence that the researcher be placed 'in the same critical plane as the overt subject matter', including the class, race, and gender of the researcher, and the researcher's beliefs and actions:

> Thus the research appears to us not as an invisible, anonymous voice of authority, but as a real, historical individual with concrete, specific desires and interests . . . This requirement is no idle attempt to 'do good' by the standards of imagined critics in classes, races, cultures (or of a gender) other than that of the researcher. Instead, it is a response to the recognition that the cultural beliefs and behaviors of feminist researchers shape the results of their analyses no less than do those of sexist and androcentric researchers . . . This evidence too must be open to critical scrutiny no less than what is traditionally defined as relevant evidence. (Harding 1987: 9)

In the study I describe here, I sought to combine conversations in a naturally occurring setting and a longitudinal study with the more common device of the group interview. I saw this as a way of placing in jeopardy my own research perspective and assumptions—by discussing my observations and findings with the women in the study, thus allowing for more complex analyses to emerge.

Lesley and Wade

Lesley and Wade live in a small two-bedroom apartment downtown above Wade's musical instrument repair business. Lesley works five days a week as a dog-groomer. They have a lot of worries over money and cannot afford the same kind of housing, furnishings, or appliances that other families in the group enjoy. Since the first meeting, when the babies were only three months old, Lesley has been consistently more open, even confessional about problems she has with her daughter Kelly: losing her temper, feeling tired, handling tantrums. Lesley often talks about reading books about child-rearing (or seeing child experts on television), which she uses as a source of inspiration and solace, and offers the tips she has gleaned from them to others. She does not want to hit Kelly, and she is more open than any other mother in the group about the temptation to do so. Lesley is very dedicated to the group: she volunteered to be a co-ordinator, and regularly phones around to everyone to announce the next meeting time and location.

Lesley offered up an extremely negative opinion of television viewing as an addiction, as a mindless, passive activity—*and* was simultaneously very frank about television viewing being something she very much likes to do. Wade was usually silent in the group, as though unfamiliar and uncomfortable with the middle-class codes which sanction a masculine display of emotional openness and self-revelation. During our two-hour discussion he spoke up only a few times, and got the floor only once, to say that since the spread of television no one plays musical instruments any more, because you can't play and watch television at the same time. Wade's disapproval of television is one source of Lesley's extremely critical view of her own television watching.

When Kelly was two, she received a videotape of *Bambi* as a Christmas gift. Twice during that winter Lesley took me aside to confide, albeit hesitantly, how many times Kelly was able to watch *Bambi*. I assured her that my daughter did the same thing with many videos, and Lesley said she thought it was good that Kelly verbalized a lot while she was watching it. Lesley felt a responsibility to filter the material Kelly watched on television, but she approved of *Bambi* because it was a story about animals and saving the forest. In June 1990, when I taped the first group discussion, Lesley expressed many more negative opinions about television's effects. She had a great fear that she must curb her own viewing so that she did not teach Kelly by example how to be a television zombie:

> It's where I go to lose my mind when I come home from work and I don't want to think and I don't want to do and I don't want to anything. And yet I don't want Kelly [*emphatically, but leaving the sentence incomplete*]. So I'm really weaning myself off TV lately. I see all these ads for candy and cars and ridiculous looking dolls and just junky stuff. I want Kelly to want to be entertaining herself.

Lesley was especially troubled at that time by her addiction to the daily sitcoms shown five days a week beginning at 7.00 in the evening. Before Kelly was born she never worried about watching these shows, but now she worries a lot. Lesley, who works a physically demanding job and then comes home to prepare dinner and clean up, was the only mother who offered up a picture of her *own* television viewing. The picture was one that is usually associated with masculine modes of viewing: collapsing after work in a chair in front of the television screen, beer in hand. At this stage, Lesley was the only mother who expressed a strong desire to watch TV herself, although she said she was terribly troubled by it.

Shortly before the first interview, Lesley changed daycare arrangements for Kelly. Kelly had been to a homecare situation, which Lesley really liked in every way except one: the television was on all the time, and this worried her a great deal. The new arrangement was much more expensive and less convenient, but it offered 'activities' and there was no television. Lesley was proud to report that within a month after the change Kelly was no longer using her finger as a gun, or talking about what she had seen on TV during the day.

Lesley herself talks about television every day with the women at work. 'That's all we talk about', she said. 'We all work in one big room and every day it's, did you see . . . ?' But Lesley doesn't want to resemble these women and she doesn't want Kelly to grow up to be like them. Every other adult in the group, when asked about it directly, said that television never comes up as a topic for discussion at work, 'except maybe a PBS [Public Broadcasting Service] special or something', one woman added, a statement that is doubtful in accuracy, but significant as a representation.

Lesley differs from the group in other ways. On a test that a visiting psychologist gave the group to help us sort out our priorities as parents, Lesley marked 'spirituality' as most important. I had marked intelligence, and others marked similar character traits. After describing how she disapproves of the kinds of toys advertised on TV, Lesley went on proudly to recount a story about some of Kelly's play with horses that she had observed:

> I bought a farm set for Kelly and out of all the farm animals she's picked out the horses (thank goodness I love horses) and she makes all kinds of wild scenarios, I mean she's already starting to tell stories. She had a horse going 'help help help!', and took another horse over to rescue, and made little conversations with the horses and stuff. She's already got this concept of helping each other, [*slowly with emphasis*] of coming to aid.

For Lesley, the point of the story was that Kelly was acting out a story of helping, not that she was being clever, or exhibiting her language skills—the point of most of the stories that the more affluent parents tell.

Carla and Ron

Carla is the co-coordinator of the group, with Lesley. She and her husband Ron live in a large home built in the 1920s, and they drive a BMW and a Volvo. Ron started his own business in vitamin and dietary supplements, selling such popular alternative medications as zinc lozenges. Carla and Ron now have two sons, Douglas, a toddler, and Jake, still a baby. He is Jewish, she is Catholic and the only full-time housewife in the group (they are the only ones who can afford it). Their son Doug was one of the last in the group to attend daycare, and he is now enrolled in a pre-school about ten hours a week. Both Carla and Ron moved to the Northwest from New Jersey in the late 1970s: they were hippies then—something they now laugh about.

Ron and Carla have the most lenient, permissive attitude towards television in the group. Carla laughs that they have fifteen hours of *Chip-n-Dale Rescue Rangers* on tape—with the commercials zapped out, Ron adds. Carla tells humorous stories about trying to prevent Doug's attention to television from wandering, especially now that she has the new baby to take care of; she talks to him about the cartoons, trying to get him interested again. Carla is charmed, not frightened, by the influence of television on Doug: he regularly has 'cheese attacks' like his favourite cartoon characters, and he goes to the refrigerator and serves himself.

When one mother complained about violence on Saturday morning TV, reporting on it as though she assumed most of us would be unfamiliar with it, *Ghostbusters* came up as an example. Carla broke in to report that it was one of Doug's favourites:

CARLA: We watch *Ghostbusters* and we watch *Slimer [and the Real Ghostbusters]*. He hasn't noticed the *Ghostbuster* toys yet, I'm waiting for that shoe to drop [*laughs*]. What I feel reassured about is he was definitely exhibiting a fear of ghosts and monsters—I mean, really severely afraid of them—but now he's into shooting them with fire.

RON: Through the media he has found a way to deal with them.

PAULA [*laughing*]: Yeah now it's shoot mom with the fire when he's angry at me.

Lesley had said adamantly that 'she won't tempt Kelly' by taking her to stores like Toys-R-Us, and that she is 'insulted' by the commercials for toys. In contrast, Carla goes shopping with Doug several times a week for amusement, using the store and its toys on display as a playground. Ron interprets Doug's watching commercials and asking for toys as a positive indication of his developing communication skills: an active process. Because Ron is himself an entrepreneur, the transaction of advertising and consumption is untainted.

RON: One thing I think is good is that Douglas is understanding that there are choices out there. It opens some form of communication. He knows that it exists and he communicates it to Carla that he wants it.

For Ron and Carla, everything Douglas does is a sign of his intelligence, so they are able extend this to his television viewing as well. Doug's life chances, his economic security, and his cultural capital are so secure that his parents can interpret everything he does—including watching television—in relation to the dominant version of childhood as active. (Ron and Carla are using a strategy very similar to the one cultural studies people use to defend the popular audience.)

By contrast, Kelly's future is less assured, and her TV viewing is subject to the negative interpretations of manipulation rather than learning, passivity rather than activity, victimization rather than choice. Lesley essentially employs a moral code to interpret Kelly's behaviour and the consumer culture which surrounds her. At the toy store, Doug is making choices and communicating; Kelly is getting tempted. Doug is using TV to conquer negative psychological states (a uses and gratifications position); Kelly is getting brainwashed, imitating anti-social values (the bullet model of effects).

In a conversation about styles of play, Ron proudly told the group: 'Douglas's new toy is the computer mouse, he's totally jazzed. He was

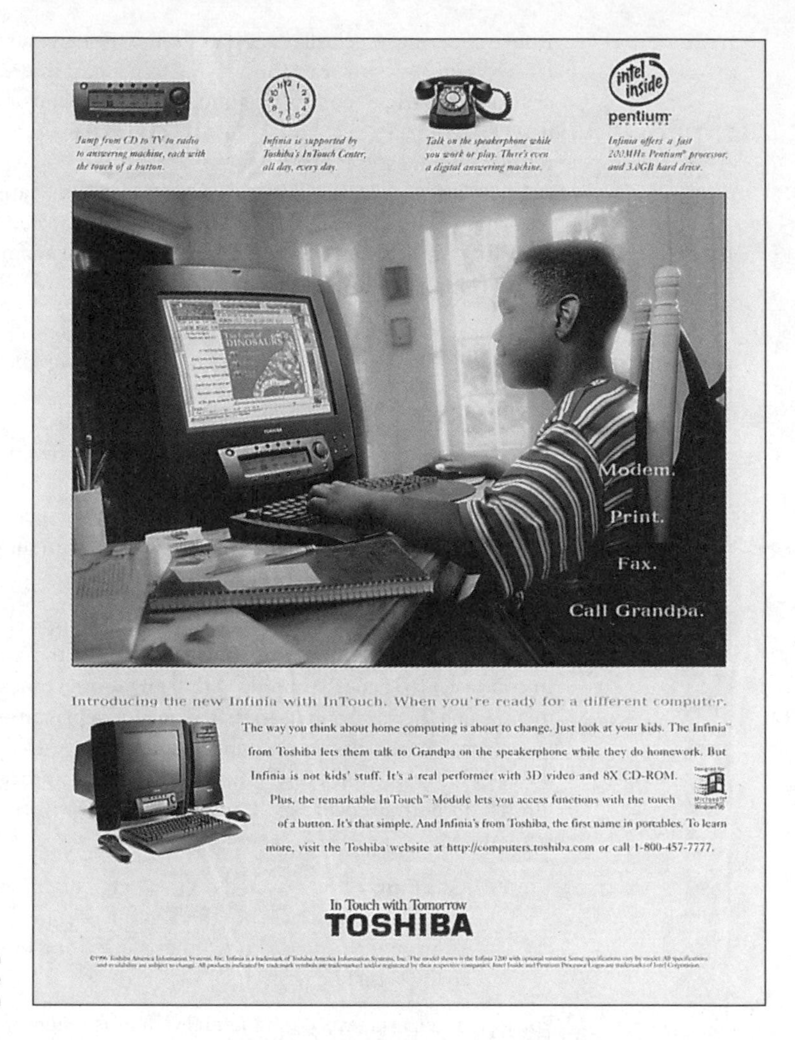

The charmed image of a boy in front of the computer screen

actually manipulating and using it!' If the television is commonly considered the bad screen for children because it causes passivity, the computer is the good screen because it is construed as active and intellectual. Thus, public libraries virtually never house television sets which can be used by their patrons: the presence of a television in the children's section of the library would be widely offensive. But computer screens are available in many children's libraries and their use is enthusiastically encouraged by parents and librarians. There is no problem in viewing Doug, a boy in front of the computer screen at the age of three, as the author of his own intentions.

Different things are at stake in terms of what Kelly and Doug know about television. Kelly's parents are aspiring to be middle class; Doug's parents have very much already arrived. Lesley's fears about Kelly's viewing were confirmed in the midst of the group discussion when the two mothers who are teachers stated unequivocally that children who watch the most TV do the worst in school. (They also reported that teachers never talk about television with their colleagues.)

Lesley is right in a way to worry about how much television Kelly watches if this will be taken as a bad sign at school, and Kelly might indeed get off to a more precarious start than Doug, who will be extremely verbal (having had adults at home with time to listen to him), already acquainted with the computer, and filled with the expectation that his desires are important ones.

Ed and Laura

Ed and Laura live in the university neighbourhood, in a 1920s home (like Ron and Carla's) which they renovated themselves. Ed is a tax accountant and Laura works half-time in a job-sharing situation as a grade-schoolteacher. Unlike Carla and Lesley, Laura had never spoken to me about the media before, although she has repeatedly asked me about my experiences working at the university. During the taped discussion, Laura's participation consisted of comments indicating (announcing?) her total lack of familiarity with Toys-R-Us, with Saturday morning television, and with commercials. Ed said that they prefer going to the park to watching television. When Laura first got the floor, she told a story about the horrible toys that children in her classroom bring to show and tell. Laura finds many toys disgusting and is annoyed at the disruptions caused by their presence in the classroom. As a grade-schoolteacher, she is convinced of the terrible effects of television. (The other mother in the group, who is a high school English teacher, has moved the television set out of her family's living-room, to prevent the son from watching it at all.) Despite the fact that her daughter Victoria only watches PBS, Laura, trained as a teacher and horrified by television, feels deeply that even this is wrong.

LAURA: I go to work at 11.00, so between 8.00 and 11.00 I have to get ready and do a lot of things so I turn on *Mister Rogers* and *Sesame Street* and I have her sit there and watch for an hour and a half and then I feel really guilty—

ED [*interrupting*]: But she usually doesn't watch for that long—

LAURA: Yes she does. You're not there.

CARLA [*interrupting*]: What's wrong with watching *Sesame Street*?

LAURA [*crisply*]: She should be doing other things. [*Pause*] She eats her breakfast while she watches *Sesame Street* and then *Captain Kangaroo* comes on at ten, and if I let that go, you know, if I let her *continue* watching even *then* I feel horrible! So usually at ten I turn it off and I say, 'why don't you find a toy to play with?' That's after *Mister Rogers* and *Sesame Street*—an hour and a half! And she'll sit through the whole thing—depending, she gets up and down.

At this time three other parents interrupted with stories about their child's attention span for TV viewing, and the different kinds of activity that they engage in while watching TV. Several minutes later, Ed gets the floor and reasserts the negative interpretation:

ED: What bothers me is that it's totally passive, there's nothing for them to do, they just absorb what's coming to them and they don't take part in it.

There is much more discussion and a few minutes pass before Laura, rather emboldened by the fact that it has now emerged that by comparison with many other children in the group, Victoria doesn't watch that much television, goes back to the earlier question, 'What's wrong with watching *Sesame Street*?' Laura is speaking to the group, but her remarks are directed rather pointedly at Ed:

> LAURA: *Sesame Street* I think invites some participation, *Sesame Street* is not a totally passive type of show. One day they were singing in Spanish and Victoria started singing with them—
>
> ED [*interrupting*]: Well that's excellent but I think that when they develop the habit of watching a lot of television as we watch TV and it's totally passive you just sit back and it will take your mind wherever.

None of the fathers in the groups felt that monitoring television time was their personal responsibility, although many of them, like Ed and Wade, disapproved strongly of television viewing and increased the guilt that their wives felt about letting the children watch it.

Mothers are both the enforcers of television discipline and the ones who suffer from its exclusion or limitation in the household. When Laura concurs with Ed's idea that Victoria shouldn't watch television because it is passive, it means that Laura must get ready for work in the morning with Victoria at her feet. The price paid for allowing Victoria to watch television is already feeling horrible. Lesley pays perhaps the highest price for her negative view of TV: it leads her to have to change daycare arrangements, it causes more fights with Kelly, puts a strain on Lesley to entertain her daughter or to deal with her boredom, and it removes Lesley's one break during the day ('when she doesn't want to do, she doesn't want to anything'). By interpreting television viewing as passive and bad, Lesley loses the only leisure option available to her during the evening. Carla's solution—to upgrade Doug's viewing to be active—seems to be the happiest one. Unlike the other parents, Carla and Ron are united in their positive feelings about television. Because Ron works out of the house, they both have an equal stake in the distraction that television offers. Doug's viewing is almost necessitated by the fact that Carla does not work outside the home and Doug is only at pre-school a few hours a week. Carla is therefore with Doug and the new baby twenty-four hours a day. The parents have a rule that when Carla goes crazy taking care of the kids, she gets Ron out of his office to take over the childcare.

Mothers and the Leisure Gap

I turn now to a focus group discussion that I organized two years later with the mothers unaccompanied by their spouse. I was interested to explore the issues related to leisure and domestic labour more directly, and I felt that the mothers would be much freer in their discussion if no men were present. This assumption is something that I will examine at the end of the chapter.

In my prologue to the group discussion, I attempted to describe broadly both the more positive view of TV's effects on children taken by Hodge and Tripp and others, as well as the analysis of the 'leisure gap' between men and women offered by Sarah Fenstermaker Berk (1985), Ruth Schwartz Cowan (1986), and Arlie Hochschild (1989). My frame was using television programmes and videos for the children so that the mother gains time to do something else (such as chores or newspaper reading). In this session, my method fits more closely into paradigms of feminist participant-research, in which one of the goals is to share the analysis with the subjects and to do so in ways that might be helpful.

ELLEN SEITER: Let me explain what I did with what we talked about last time . . . I was really interested in what a lot of you said (and what I myself felt was this real tension between kids watching television as a pretty convenient thing and on the other hand, this constant anxiety about the negative impact that it's having on kids) . . . It turns out from the reading I've done about it, that it's actually pretty doubtful that watching television does have a bad effect in the way we usually worry about . . . I've done a lot of reading about why everybody, why all of us, appear to be so pressed for time. Why do you need television to get time? There seemed to be a real time crunch . . . You're doing laundry every day instead of once a week, and you're shopping at more places for more different things and you're driving around in the car more, and this leads to what they call a leisure gap. That is, the worst thing that happens to women in terms of having time to themselves is having children [*big laughs*]. So I'm interested in whether there's any time to watch television.

Lesley was the first to speak, eagerly, about her guilt over viewing, her difference from the others at her job, and the tensions she felt she was causing by watching this way.

LESLEY: I have been doing so-o-o-o-o much thinking about this lately. Every night I come home and flip on the news—this is my justification for watching TV. I flip on the news, but I'm in the kitchen, so I can hardly hear it and I can't see it. Then, my junk show is *Star Trek*. So, that's seven [o'clock] to eight so I figure that's two, two and a half hours of television. So that's not like this maxed out [*US slang: added up to*] fifty hours of TV a week that they talk about, you know, with everybody going into this ozone.

LAURA: Who has time to do that, when do they do that?

LESLEY: All night long and all weekend they just sit in front of the TV and watch sports shows . . . and they're watching all kinds of sitcom stuff or, you know, whatever.

ELLEN: Do Wade and Kelly let you watch it?

LESLEY: Wade hates it. He disappears downstairs to work. So our communication has gone completely to zero. I mean we don't talk about anything anymore and it's really bad.

LAURA: Maybe you should turn off *Star Trek*!

LESLEY: But it's hard. It's an addiction. It really is. And this is what I'm combating right now because it's completely eliminated my self-discipline.

When Lesley spoke up, Laura was quick to come in as an interrogator, encouraging Lesley to see her situation as one that she should feel guilty about, and insisting on the need for her to take self-disciplined action. Despite the fact that Laura had once challenged Ed on this very topic during one of our discussions, in the absence of the fathers, Laura took over the same role as her husband in previous discussions: harshly critical of television viewing as passive, bad, and destructive of family life. Despite the fact that other women in the group—especially Carla, Tess, and myself—form a chorus of sympathetic voices, attempting to 'grant permission' to Lesley's behaviour, Laura's is the voice that Lesley responds to most directly, and most defensively.

CARLA: But do you have to deal with Kelly through this? What's Kelly doing? You can follow the story line—

LESLEY: By the time *Star Trek* comes on dinner's cooked. So that's the other thing—

ELLEN: Because it's in a different place?

LESLEY: No it's right there, with the living-room and dining-room table right there next to each other . . . six feet away. But her attention span isn't on her food or concentrating on eating so she doesn't eat dinner then she wants dinner at eight or nine o'clock.

LAURA: Why don't you tape it and watch it late at night?

LESLEY: No, that's the thing I shouldn't. I need to stop watching TV altogether. My self-esteem has gone way down because my self-discipline . . . the TV has sapped my ability to structure my own time.

TESS: I don't think so. You need a break, a reward.

LAURA: But you can't have it on when you eat dinner.

ELLEN: We always have it on.

LESLEY: I've been talking about it with my mom and she's . . . English and reading and all that is her forte—I think I said that right—it's her forte. She agreed with me wholeheartedly, her husband died a few years ago and she's been watching a lot more TV and she said 'I have no desire to read anymore it's just so easy to go in front of the TV.' And when you do that enough it really saps the energy to pick up a book and sustain the concentration.

ELLEN: Do you think you're kind of tired, though, when you're starting to watch *Star Trek* or do you think *Star Trek* is making you tired?

LESLEY: Well I'm tired after a full day's work because I'm standing there and we don't get this nice fifteen minute break in the morning, fifteen minute break in the afternoon and half hour lunch in the middle. We're booking from eight to five straight through. And I come home and I just flip on the TV because it's so easy. But it's not constructive use of my time and I've found that I, that my discipline level has gone way down—

TESS [*with disbelief*]: Discipline level?

LESLEY: Discipline level in just keeping the TV off, just cooking dinner, and you know, that's just kind of my junk show—

LAURA [*speaking over Lesley, but no one picking up on it*]: We watch the *MacNeil-Lehrer Report*—

LESLEY: Because I don't have any desires to do anything constructive like spend time with Kelly and read a book or sing songs or . . . maybe if I weren't watching TV I would have more energy . . . I—

TESS: You know, I think you hit points in your life where you go, I do anyway, where I'll watch *The Tonight Show*, you know, and be up until 12.30, and then I'll start watching *Letterman* and I enjoy it, because it's like, it's almost a form a rebellion, where, I don't have any leisure time, so I'm just gonna take it, you know! And if it means it's going to make you really tired, to me, it's worth it. It's worth the trade-off because I feel like sitting there watching *Letterman*, and I still get his jokes, and I feel like 'I'm not dead!' I still understand adult things, I can flow with that subtlety. I just go through phases like that.

LESLEY: But see what I was thinking was I would probably be much better informed if I spent that time reading the newspaper instead of watching TV—

TESS: But what do you want to do? How many have-tos are you already doing?

LESLEY: I want to have communication with Wade. I wanna you know . . . He disappears because the TV's on then something's got to change.

ELLEN: He disapproves of you watching it?

LESLEY [*emphatically*]: Oh, yeah.

ELLEN: TV in general or just *Star Trek*?

LESLEY: TV in general.

CARLA: So maybe you don't want to talk to Wade.
[*Big laughs all around*]

I offer this excerpt as an example both of 'realistic' support group talk—the kind of conversation that happens about television among people who know one another fairly well; and as symptom of the discursive limits of talk about television. It reveals what Lesley can say about television and how she feels about it; but also what she cannot say about it. Concerns about television viewing were expressed differently at different times, and contradictory statements were often made in the course of one meeting. All the women in the group were likely to change the tone of their statements radically after a remark from someone else, but the stakes—of self-esteem, of role-expectations for being a good mother, and a good wife—tended to remain the same.

In my preamble to the group discussion, I had unwittingly positioned myself in the type of middle-class norm represented by Laura—unconcerned about the poor example-setting of watching television myself. Because I had completely given up on choosing my own television programmes, already deferred to my children and husband in choosing television programmes, I had written it off as a leisure option myself. The three women—Tess, Carla, and I—who most encouraged Lesley to forget the guilt and enjoy her 'junk show', were all women who had managed to work reading the newspaper into our daily routines, even if this time was jealously guarded. Tess advocated vigorously, in this discussion and others, the importance of trying to take it easy. In fact, Tess had it somewhat easier than Lesley: she worked part time in an office (rather than full time doing physically demanding work) and she and her husband could afford a big house and household help. My own position as a questioner is different from any of the other women's, though no less motivated. At times, my attempt to elicit detailed descriptions of the pattern of Lesley's TV viewing forced me to postpone the more supportive comments of Tess and Carla.

In part, Laura's greater authority in the group—her symbolic capital, to use Bourdieu's term—as a schoolteacher helped to push Lesley from a description of her viewing as an understandable release at the end of the day to the metaphor of physical addiction. Laura took a year's leave after the birth of her first daughter, then returned to work as a fourth-grade teacher. She stayed home for two years after the birth of her second daughter Angela (something she found very difficult to do), then got a new half-time job as a foreign-language teacher at the high school (in the same working-class community Lesley worked in). Laura is both harshly critical of Lesley's TV dilemma, and reports that she continues to use the television set as a baby-sitter in precisely the way she did two years earlier—despite her husband's objections. Her critique of commercial television is focused more centrally on the irritations caused by the commercials, and children's demands for advertised products. In fact, she seems to have grown more resentful of her lack of leisure time (her harsh judgement of Lesley being tinged, perhaps, with jealousy) and of her husband's lack of participation in child-rearing.

LAURA: Have you seen the new show, *Lamb Chop*, yet?

SEVERAL: Yes [*approving murmurs*] Shari Lewis. Oh, I love her.

LAURA: Well, I feel bad that I sometimes turn *Lamb Chop* on at eight, and it's still on at *Captain Kangaroo*, and if I don't turn on the TV in the morning, Victoria will start playing with her dollhouse, do all kinds of interesting, neat things, but sometimes I want to read the paper for breakfast, and I don't want to be interrupted, so I turn on *Lamb Chop*, and she sits there and while she's eating breakfast and while Angela is there, they watch *Lamb Chop*, they watch *Mister Rogers*, *Sesame Street*, *Captain Kangaroo*. This is not every day. We [talk]. There are periods when, if I want to get a lot done, then I turn that on, but I do not ever turn on the network TV. Sometimes she does, and those ads start coming on, and she's never seen half these things, since we don't go to toy stores. And as soon as she sees something, she starts yelling 'MOMMY! MOMMY! COME OVER HERE FAST! GIMMEE THAT! GIMMEE THAT! GIMMEE THAT!' And I can't deal with it. So I just never turn that on ever, so she never watches cartoons.

A few minutes later Laura turns to the topic of her own leisure, still carefully establishing her restriction to the quality fare of PBS, even if she is hopeless about accomplishing the activity of television viewing. Like many of the women interviewed by Morley for *Family Television* (1986), Laura has difficulty relaxing in front of the television set. Her resentment towards her husband is much more openly expressed than it was in the group interview; she also feels the lack of relaxation more acutely now that she has two children and is employed outside the home again.

LAURA:	Do you ever listen to the radio? We always listen to *Morning Edition* and—what is it called? The evening NPR [National Public Radio]? [*Talking—unintelligible*] And then, after dinner, while we're getting the girls ready for bed we'll watch *MacNeil-Lehrer*, snatches of it, while we're getting them ready for bed. I always am doing something else while I'm watching TV, I can't just sit and watch TV. I have laundry to fold, or I have something to do ['*Yes!' someone cuts in*], or what bugs me is that Ed can just sit there and do nothing and watch TV, and it really irritates me. He should be able to fold laundry while he's watching TV. I have to be doing something, because there's so much to do. But I guess, he works really hard during the day, and I guess he thinks he needs to just relax in front of the TV, but I can't just sit in front of the TV and relax. I have to be doing something. It's hard, even when we rent a movie, for me to just sit and watch the movie.
SOMEONE:	Yes.
TESS:	That's the only time I can really be a couch potato.
LAURA:	And I don't have any favourite shows anymore. I had one show that I watched, and that was *Thirtysomething*. And they took it off, and now I have no shows.
SEVERAL TOGETHER:	*Northern Exposure*? It's on Monday nights at 10, and it's the only one—
LAURA:	Well, I heard there was one called *Fly Away* or something—
SOMEONE:	*I'll Fly Away*—it's very good.
LAURA:	But I'm always busy between 8 and 8.30. I take Angela upstairs and rock her to sleep.
TESS:	We're VCR junkies—we record it and speed it up.
LAURA:	We tried to record it one night, and our VCR wouldn't get that channel. It wouldn't programme in. We don't have cable.

Laura's statement resembles Lesley's in its pessimism about everyday life, its air of complaint, of all factors conspiring against her ability to enjoy herself. Laura's disapproval of Lesley is cast in a new light when we consider how strongly she is cast in the role of a drudge: no possibility for entertainment, a never-ending set of chores awaiting her. Clearly, Laura is sceptical of the position that men work hard all day and need their time free ('I guess he thinks he needs to relax—'); she implies that she works harder than he and never has the time to relax, so doesn't 'need' to. Even if her husband and children allowed her to watch TV (and every media experience seems to be in the company of her husband), 'they' (the networks) might take her favourite show off the air. While, in the earlier interview, Laura was somewhat resentful of her husband's censure of letting the children watch PBS, in the focus group discussions she consistently adopts a studied ignorance about television as a sign of distinction, and she herself disapproves of watching TV, offering NPR's news programming as a healthy alternative. It is only after much discussion that Laura comes around to voicing some regret about no longer having even a single show to watch.

Group discussions have a life of their own. That is, it is impossible for the interviewer to control them, which is one reason they yield interesting results. Despite my attempts to communicate to the group the type of permissive attitude towards television viewing popular among cultural studies academics (I boldly claim, for example, that we always have the television on during dinner), what this research has been perhaps most useful for is documenting the ways that admissions of television viewing—or any other kind of less-than-ideal parenting behaviour—run the risk of drawing censure from other women in the group, as much as they have the potential to garner support. I tried to steer Lesley towards considering the level of fatigue and obligation she was shouldering, as when I asked if she was already tired before watching *Star Trek*, but Laura had only to give her a little push towards a derogatory interpretation—by mentioning public television, viewing together with her husband, the 'just say no!' approach of turning the set off—to switch Lesley into an attack on her own self-worth.

Our group involved a great deal of implicit competitiveness and one-upmanship. Class issues were suppressed (we're all mothers together), and though the difficulty of carrying out the ideal project of middle-class child-rearing was continually addressed, we rarely spoke about how much more difficult that project was for some than for others. None of the other mothers worked a nine-to-five job, as Lesley did, and Lesley was the only other mother besides me working full time. None of us had jobs demanding the type of physical exertion that Lesley's did. As a caste-like group, we were familiar with a long list of shared problems: the five o'clock 'arsenic hour' with children, the seemingly endless burden of household chores, the demands of young children for attention, the wife's burden to ensure proper communication and closeness within the marriage, the leisure gap. But the ability to free oneself from the 'have-tos' was differentially distributed along strict lines of education and income.

Lesley's comments during these group discussions suggested a rather pessimistic view of entrapment in feelings of guilt, but my contact with her the following year reminded me of the possibilities for resilience and for change, and the possible benefits of attempting to address directly the issue of guilt over viewing with women. The last time I saw Lesley was at a going-away party at my house in the summer of 1993, shortly before I moved away from the area. Lesley had a lot of news: she explained that she and Wade had finally faced the crisis of the failure of her husband's business, decided to close it, and Wade had returned to the community college to learn nursing. The time of the taped discussions was the bleakest point in their marriage and they had been contemplating separation and divorce. It is interesting that in the excerpts from these discussions quoted here she attributes the stress on her relationship with Wade to her own pursuit of a little leisure-pleasure watching television—not to the economic pressures on the family, which she may have felt embarrassed to discuss. The persona that Lesley presented at that time—depressed, worried, anguished, and guilty—was a transient, passing thing. While the focus group concentrated on her TV 'addiction', there were many other things going on in Lesley's life with a more directly causal effect on her state of mind than her habit of watching *Star Trek*. Given the chance to talk to Lesley over the months that followed, I was able to capture the quite different portrait of her TV viewing that she constructed once her family crisis had passed and once she was in an individual conversation rather than a group. Lesley still referred to herself as a 'TV addict', but in a bright, cheerful way, and she was eager to discuss current shows and the American Movie Classics cable network with me. I told her that I had just been to a conference where several academics were talking about *Star Trek*. In the context of that conversation, Lesley told me brightly that she was still watching *Star Trek: The Next Generation* and that she loved the character Rykker, whom she called a heart-throb.

Lesley's reticence to discuss her financial and marital problems recalls Kamala Visweswaran's caveat that feminist ethnography cannot assume the willingness of women to talk. As Visweswaran suggests, 'one avenue left open to it is an investigation of when and why women do talk—assessing what strictures are placed on their speech, what avenues of creativity they have appropriated, what degrees of freedom they possess' (1994: 30). Lesley's resilience in defending her leisure media practices is a good example of this kind of creativity. The pained negotiation of her problems in the support group point to the strictures that she experiences in her struggle to meet the image of the good mother constructed by advice literature, in her striving as a working-class woman in a middle-class support group, and in her claiming of leisure time in the face of her husband's criticism and emotional withdrawal.

Because my research documents the talk in a self-help group, much of what I recorded were statements explicitly designed to shore up the identity of each family—the 'social desirability' effect was high (Buckingham 1996). As parents, we were all conscious of making public, social presentations of ourselves and our children—just as Morley's families might have been in his interview (Morley 1986). Self-help groups are, in a way, devoted to making statements (and to the exchange of information—often about consumer goods and services). The group was not a forum for free speech: differences in social power outside the group determine who can speak and what can be said. For example, during our early discussion one of the fathers described with great disapproval some friends who do not turn off the television set when visited by guests and who eat meals in front of the set. Some of us in the group did practise these very things at home, but no one could confess to such behaviour after his comment. At later meetings—especially those attended only by the mothers—more frankness about differing domestic practices developed.

My own speech and participation were equally bounded by various kinds of stricture and various kinds of investment throughout the project. When I decided to use the group for research purposes I had a different kind of self-interest in seeing the group continued. As feminist researchers have noted, there is 'a usually unarticulated tension between friendships and the goal of research. The researcher's goal is always to gather information; thus the danger always exists of manipulating friendships to that end' (Acker *et al.* 1991: 141). I was always conscious of ulterior motives in the efforts I undertook (telephoning, inviting over, chatting at the grocery store) to sustain and nurture these relationships (especially after the birth of my second child in 1991, when the strains of the double day became more acute).

I have focused on class differences among the families participating in this support group, but I also noted a large degree of structural similarities among these families. Writing more than twenty years ago, Joan Acker commented on the inadequacy of traditional formulations of class in defining women's position, with their tendency to ascribe the social position of the family to the male head of household, assuming that her social status is equal to that of her partner—and assuming she has a husband. Acker suggested that

> females can be viewed as constituting caste-like grouping within social classes. Female castes, using this approach, may have certain common interests and life-patterns. In addition they may share certain disabilities and inequities. At the same time, female castes are imbedded in the class structure and each is affected by the class which envelops it. Class differences in ideology, life-chances, and life-style may obscure the identical nature of many structural factors affecting female castes. (Acker 1973: 941)

Acker's formulation seems old-fashioned, at odds with postmodern conceptions of gender, yet it highlights the ways in which raising children tends to lock down gender roles, proves the relative durability of traditional expectations of the kinds of work mothers and fathers do, and often causes people to retreat to relatively familiar, durable gender roles (Busfield 1987). Households differ in terms of how much work women do to maintain them, but the most important factor in predicting the workload is not income, or education, or employment outside the home, but the presence of children in the home (Berk 1985). Of all the types of domestic and reproductive labour assigned by social convention to women, childcare is the most strenuous. Women get less help from neighbours, relatives, or older children in doing it, and there are much higher standards for its performance than there used to be (Strasser 1982; Cowan 1986; Dholakia and Arndt 1987). The structural factor most affecting the mothers of young children in my study, and most important for the study of television viewing, was not income level *per se*, but the lack of leisure time.

On Interviewing Women

I would like to return to the question of interviewing women in the absence of men, and the reasons that I chose to segregate the mothers from the fathers as this study developed. In my previous work with soap opera viewers and in the research presented in this chapter and the next, I focus on interviews among women where a certain level of rapport was established and where the subjects were much more malleable to my direction of the interview, more forthcoming, more interesting, because they related the media to their personal histories. Compared to many interviews with men carried out during the soap opera project, the support group mothers and the pre-school teachers were 'better' subjects, and my emotional investments were stronger and more positive while I was doing this research.

My own involvement in the everyday labours of caring for small children had a certain levelling effect on the class differences between myself and the women I studied. The mothers in the support group and the daycare workers I will discuss in the next two chapters sought a common ground on which to talk to me: as a fellow teacher, as a mother, as a career woman, and we did not focus on the substantial differences in income. Going into the field to visit daycare centres and pre-schools, I felt an enormous admiration for these women, and a sense of the injustice of their wages. At times, this kind of identification nearly overrode my political differences with them, as with the fundamentalist Christian women in our study. But as I hope to have demonstrated here, interviewing women together does not mean that interactions will be routinely 'nice', will be unmarked by other dimensions of social power and status, or will be somehow democratic.

Although much influential work in audience studies has involved women interviewing women (Hobson 1982; Radway 1984; M. E. Brown 1990; Press 1991; Gillespie 1995), some consensus has emerged that, by leaving men out of the sample, the work veers towards gender essentialism. Several scholars have now called for more audience research with and about men (Allen 1989; Livingstone 1992; Ang and Hermes 1993; Schwichtenberg 1993). Liesbet van Zoonen summarizes these arguments in her textbook *Feminist Media Studies* when she explains:

> Although we have increasingly detailed insight in the use and interpretation of 'women's' genres by female audiences, we know next to nothing about the use and interpretation of 'men's' genres by male audiences . . . Although the research carried out so far does tell something about women, its theoretical flaws have prevented extensive ventures into issues such as the construction of gender discourse, the intersection of gender with other discourses, and the disciplining and regulatory effects of gender discourse on various levels. (1994: 125)

I wish now to enumerate some of the potential problems of studying men—to explain why women will outnumber men in the case studies that follow in the next two chapters.

Incorporating more men as research subjects in group interviews threatens a loss of information about women. Lesley simply could not have described her frustrations, guilt, and feelings of overload in the presence of her husband —or, probably, of any men. When men and women are interviewed together it can restrict what many women subjects will say. How much more might have been said in David Morley's *Family Television* interviews, for example, had the husbands not been present? When Jhally and Lewis (1992) turn to their analysis of *Cosby* and racial discourses in the USA, considerations of gender—either in the situation comedy itself or as a structuring factor in the group discussions—all but disappear. How would men and discourses of gender deconstruction have fitted into the *Women Viewing Violence* focus groups of survivors of domestic violence formed by Schlesinger *et al.* (1992) to study perceptions of media violence against women? Ann Gray reminds us of how often women are relied upon to report on other family members rather than talk about themselves, and offers a good rationale for her decision to exclude men and children from her study of VCRs: 'Women are far too readily seen as representatives of their families by researchers and the state alike, and my concern here is to address them as individuals occupying particular social positions, and not to lose sight of their own distinctive viewpoints' (1992: 12).

Even in small numbers, the presence of men can dramatically affect what is said in an interview. In an interesting study of fans of the British detective series *Inspector Morse*, Lyn Thomas reported that when she played her focus group tapes for others, she was criticized for dominating the discussion, and silencing the single male participant. Her discussion of the dynamics of this interview is a good example of ways that the behind-the-scenes story of an interview needs to be taken into account when looking at things like turn-taking in focus groups:

> The group dynamics in this discussion seem particularly influenced by gender difference. The fact of being the only man in the group seems to have elicited certain types of response from Jim, who seems concerned to make an impression on the others and even to obtain a dominant position. This behavior was met with opposition, at times verging on hostility, both from the group members and from myself. I certainly saw keeping Jim under control and sabotaging his attempts at dominance as an important part of my role as discussion facilitator. (1995: 14)

Thomas began to feel that she had treated her male informant unfairly. But in an interesting example of the benefit of some quantification, she later found that the man had spoken second most often in the group and had introduced more new subjects than any of the informants. Communications scholar Dale Spender has argued that whenever women speak more than 30 per cent of the time they are perceived as dominating the discussion (Taylor *et al.* 1993).

To follow van Zoonen's suggestion of doing more audience research with men might require having more men do the research, or at least to work more frequently in mixed research teams. On the other hand, it is important to acknowledge that some of this work in audience studies could not have been carried out by male researchers. I suspect that the call for studying men has been in part a reaction against the conventional femininity inscribed in images of women—of wives and mothers—as ever-harried, inexplicably 'guilty', or curiously defiant over the small pleasures they steal from picking up a romance novel or tuning into a television soap opera. The women I have discussed here—Lesley, Carla, Laura, and Tess—seem to conform to many aspects of these stereotypes of femininity, of motherhood. Most studies with women as subjects have sketched an often poignant—but also, to feminist academics, irritatingly traditional—portrait of symptoms of the gender gap: women always doing chores at the same time as television watching, arguing with a husband over the value of reading time. The best way to deepen and complicate this picture, however, is not through interviews with male viewers, whose points of view have, after all, been well represented in theoretical paradigms and in criticism. Rather, what is needed are increased efforts to blend audience research with the historical and material analysis of domestic relations, and with the economics of unpaid (and devalued) labour in the home in comparison to the formal economy. My study is clearly bounded by concerns that preoccupied me as a white middle-class academic: careers and work, sharing of childcare and housework, gender stereotyping. These are limitations I share with most feminist media studies—with academic feminism in general, constructing:

> feminist . . . meanings . . . were fashioned around the place and concerns of white, middle class, heterosexual, western women. As stated at the outset, this mode of feminist inquiry belies power dynamics within the academy and tells us more about the social construction of feminist knowledge in Anglo-American contexts than it does about television as a world or even national practice. (Brunsdon *et al.* 1997: 13)

The glaring limitations of this type of study are those I faced at the onset, especially its homogeneity. I believe that increasing the diversity of our samples of women subjects is perhaps more important at this moment than interviewing men. While I have noted salient differences among the women in their media consumption and their feelings about this consumption, I must also remind myself that there are other domestic arrangements—especially single-parent households or lesbian households—and other attitudes (including those uninformed by the dominant themes in advice literature) that were not represented within this small group.We need comparative work that looks at differences between single women living alone or with children and women living in more conventional families to explore the relationship between gender identity and media consumption.

Audience researchers need to attend closely to the self-selection elements in qualitative studies: they often exclude people who will not have the time, or the trust, to participate in such research. This is one reason that white media researchers repeatedly wind up with all-white samples; my support group is no different in this respect. It is likely that the self-selection factor in joining the group excluded some women of colour who tend to have little free time for such activities and may be more suspicious than whites of the interventions of social workers (Cannon *et al.* 1991: 113). Historically, social service agencies—and academic sociologists—have incurred much resentment in the black community, for example, leaving many African-American women less than eager to volunteer themselves or their families for further surveillance.

There is a powerful way in which women's experience—especially the perspectives of women who do not happen to be white and middle class

—continues to be marginalized in social science discourse, which makes arguments such as van Zoonen's seem premature. Sociologist Dorothy Smith articulates the argument—familiar as a rationale for women's studies programmes:

> [T]he worlds opened up by speaking from the standpoint of women have not been and are not on a basis of equality with the objectified bodies of knowledge that have constituted and expressed the standpoint of men. The worlds of men have had, and still have, an authority over the worlds that are traditionally women's and still are predominantly women's—the worlds of household, children, and neighborhood. And though women do not inhabit only these worlds, for the vast majority of women they are the primary ground of our lives, shaping the course of our lives and our participation in other relations. (1987: 13)

As Smith reminds us, the methods, research practices, conceptual models and theories of sociology have been built within a male social universe, and thus have not only excluded questions related to women's experience but actively participated in building a 'society's self consciousness' in which women's experience is denied, made invisible, alienated. It is one of the singular achievements of media studies and audience studies—one that deserves careful guarding—that women's experiences have occupied such a central place in the discourse.

Epilogue

What I have presented here are some of the ways in which attitudes towards television viewing emerge from particular material conditions and family configurations, and the ways in which statements about television are produced under particular types of constraint by individuals with a range of investments and defences at work. This type of design was ideal for capturing *talk* about television; to study the practices of television viewing it would be necessary to bring other types of method, such as diaries and observation, to bear on the study. I felt that the introduction of such methods in this case would have had a chilling effect on my rapport with the interviewees, casting me in the role of surveillante, checking up on their self-reports of media habits. However, I wanted to give them one last say on the subject, and an opportunity to write about it rather than converse with me about it.

Five years after our first meeting, I moved to a city some 2,000 miles away. From there, I wrote back to all the women asking them if they thought their feelings towards television had changed since their children were born, how they had felt about our discussions about television, and whether they had anything to add. Laura expressed a broad-ranging contempt for the pleasures of popular television and cast herself as a strict parent: 'no Disney videos—I think they stereotype people too much and they are also too scary'; 'I don't allow my girls to watch any cartoons. They are too violent and stupid!'; 'I don't like the girls to watch Network TV (only PBS) because of the ads—which just create these ridiculous desires for things they don't need, but they think they must have.' Most significantly, Laura seized the opportunity to convey thoughts she had had during our group discussions that she was unable to say out loud: 'Often people make references to TV shows and I have no idea what they are talking about—but I don't mind—I'm just shocked by how much TV most people watch.'

Carla expressed an increasing concern over TV and consumer culture's impact on boys, something that is moderating her relatively open and anxiety-free position about television. 'I loved TV as a kid, teenager, young adult and still think TV has a lot to offer. However, as the boys get older, and I'm aware of their growing awareness, the more nervous I am about violence and sex.' She wrote that her 'two boys spend the next hour hitting each other or doing Ninja kicks' after seeing *Teenage Mutant Ninja Turtles*. Her concern about masculinity extends to her choice of toys: 'It's actually getting to the point that I view all "boys' " toys as a bad choice.' It was interesting to see the strongly held, vastly varying opinions on the subject. 'It's interesting to realize that my opinion or position on TV is moving closer to those I felt were more conservative than I.'

Lesley wrote that she had really enjoyed our discussions. At the top of her list of what is not OK for kids to watch is the sarcastic, hurtful, put-down 'humour' on sitcoms—the very shows she was 'addicted' to when we began our meetings years before. Echoing Laura's sentiments, she writes, 'TVs turn kids into nags on poor, harassed mothers to accumulate life's most unnecessary stuff.' Lesley also added something that she never expressed out loud in the group discussions: 'I was surprised to learn most use TV as a babysitter. I don't like doing that. I encourage her to get books or paint (but I don't have two kids).' Lesley takes a little revenge against Laura here, and, contrary to the deprecating tone of the group interview, reveals a quite different strategy for evaluating her television watching as superior to that of mothers who use children's television shows as a babysitter.

On the subject of which shows Kelly watches, Lesley wrote: 'I find I have become more conservative about TV programs. I used to sit in front of just about anything but now I don't want to set that example for Kelly. I would rather, now, have our own collection of good movies and limit watching to an occasional movie.' On her list of shows that are okay for Kelly to watch, her own favourite programme—*Star Trek: The Next Generation*—is slipped in between nature documentaries and *Sesame Street*.

4

Lay Theories of Media Effects: Power Rangers at Pre-school

COMPARE THE FOLLOWING classroom scenarios. At a suburban Montessori school, the teacher bans popular videos or films from the classroom on the grounds that they promote passivity. Viewing is rare and is confined to 16 mm films borrowed from the children's library or *National Geographic* specials. Talk about television or play involving popular characters—even on the playground—is expressly forbidden. At a nearby developmental learning centre, a male teacher embraces the school's explicit commitment to a non-violent and non-sexist curriculum, disapproving equally of female narcissism as represented by Barbie, and male violence as represented by war toys. He cringes, however, at the idea that he would be perceived as the 'morality police' with parents or with children. Across town, at a low income daycare centre, kids watch network television programmes, and freely play rowdy *Power Ranger* games. Along with their care-giver, the children celebrate character theme days, where they dress in purple for *Barney and Friends* or enact plays of *Beauty and the Beast*. At a daycare centre for hospital employees, the teacher helps students make Power Ranger helmets out of papier mâché and supplies sheriff's badges and a 'jail' for Wild West play scenes.

These case studies represent a range of policies, practices, and concerns about media in the pre-school environment; they include the most affluent and the poorest centre, the most inclusive media environment, and the most restricted. I wish to present these teachers' stories in some detail in order to explore the factors that influence beliefs about media effects, and to call attention to the ways that classrooms and daycare centres are an important site for early socialization about the media and about distinctions between legitimate and illegitimate forms of cultural capital.

Lay Theories

In the case studies that follow I trace the variety of teachers' beliefs about media effects—an example of what social scientists call lay theories—and how these beliefs impact media use in the pre-school environment. My motivation in initiating this study was to compare the 'weak' theory of media effects held by cultural studies academics, by many industry professionals, and proposed in my own work on children's television (Seiter 1993) with the theories of 'strong' effects that were often implicit in conversations I had had with teachers and childcare professionals. Studying 'lay theories' allows us to compare academic and lay theories, implicating both in a relationship of mutual influence and finding contradictions in both forms of theorizing.

Going into this study I was conscious of—indeed compelled to reconsider—inconsistencies in my own position as a parent and a teacher *vis-à-vis* the media. Attending film school in the 1970s, I had been greatly influenced by the ideological analysis of popular media. Yet, as a television audience researcher and later as a parent, I was equally influenced by academic work that celebrates popular pleasure and the possibilities for expressing social resistance through fandom (L. Lewis 1992). As a mother, I was predisposed to feel most sympathetic towards teachers in my study whose lay theories suggested weak media effects (and a permissive attitude towards children) but a race- or gender- or class-conscious critique of media representations.

In engaging in the interviews for this study, I was interested to hear what teachers with a vastly greater store of direct observation than mine thought about the media. It seemed important to note that so many people who dealt with children on a regular basis held a theory of stronger effects than the academics such as myself who had much less direct contact with children other than their own, or contact only with research subjects encountered for a short time.

As used within sociology and anthropology, 'lay theories' seek to investigate the cultural determinants on people's common sense and worldview. This kind of study 'looks specifically at the sort of information that people select and reject, and how they use this information to examine or test, various hypotheses that they may wish for or have been asked to verify' (Furnham 1988: 46). Social scientists have found that there is a tendency to rely on individual, psychological explanations rather than societal or structural ones. Additionally, 'because they are rarely, if ever, presented formally, lay theories are frequently ambiguous, incoherent and inconsistent. That is, people can hold two mutually incompatible or contradictory ideas or beliefs at the same time and not be particularly troubled by that inconsistency' (Furnham 1988: 3). There is a tremendous variation in the quantity and quality of theories that people hold—teachers, as a group, are likely to hold more elaborate lay theories about media effects on children because of their many opportunities to observe large numbers of children and their closeness to scholarly discourse. Adults working in other kinds of fields would, of course, be likely to have undeveloped or non-existent lay theories about media effects on young children.

To some extent, these interviews testify to the diffusion of a theory of direct media effects—especially among the most professional caretakers. They also indicate, however, the presence of a more flexible, 'forgiving' theory of media effects—the cultural studies version, if you will—which attributes greater agency to the children, and places more value on the pleasures of popular culture. The interviews also indicate the ways that theories can be challenged by direct experience, and by the intervention of other factors. As I will demonstrate, the media are deemed most powerful by those working and living in situations of relative privilege; in the poorest centre the media are seen as only one factor—less significant than the part played by poverty, by parental absence, and by violence.

The interviews carried out for this study were shaped by the fact that we were academic specialists in the media, eliciting talk in an artificial situation about beliefs about media effects and classroom practices relating to media. Furnham notes that conducting interviews about lay theories is particularly vulnerable to some of the problems—response sets—with interview data: 'social desirability bias, faking good or bad, acquiescing with their perceived demands of the researcher' (1988: 17). Because I felt that these issues would be so powerfully determining, I opened each interview with a statement about my own theories about media effects—that I felt the dominant view often overstated or exaggerated television's effects on children. I felt this was necessary both to grant permission to those who wished to articulate a theory of weaker effects, and because it seemed only honest and fair to 'lay my cards on the table' for those who, I guessed, believed in stronger effects than I did. At the time these interviews were conducted my book on children's television and toys was available at bookstores, and I had appeared on local television. During the interviews, I had the impression that two of the teachers intended to disabuse me of my permissive attitudes towards popular culture; the others, with whom I felt an easier rapport, were visibly relieved to meet an academic—and a parent—who did not seem to stand in judgement. During each interview, we moved between our positions as professionals, as tellers of childhood experiences (much on my mind was the fact that my own mother had forbidden Barbies, for example), and as parents—and in each of these registers the degree of or potential for criticism varied. I was at my most judgemental and defensive as a parent, conscious of the fact that my own practices might meet disapproval had my children been in the class of the teacher to whom I was speaking. There are tensions throughout the interviews between my desire to elicit a full statement of the basis for rules regarding media use that might have been heretofore unarticulated and the fact that I worked hardest to elicit statements of this nature from those with whom I was in most disagreement. By paying the teachers for the interview and adopting the role of an interested listener, I attempted to honour the work that these women did and value their experience—but I was unable to hold uniformly to this attitude when I perceived certain contradictory values at work, or when I pitied children who I imagined were reprimanded in ways they did not fully comprehend.

Television and Teachers

My second focus in choosing these interviews is to illustrate ways that rules about television and toys emerge from the intersection of the material circumstances of the care setting, and the beliefs and experiences of the teachers. In the case studies that follow, I will suggest the ways that some of the work of teachers relates to status production and is explicitly tied to the censorship of popular media materials in the classroom. I will also ask how rules about TV in childcare settings help produce status differences and work to stratify and segment the childcare market.

Babysitting young children is one of the things television does best. Television is undeniably handy for calming children down, confining them to one area, reducing noise in the classroom, and postponing demands for adult attention. But such uses of television are widely condemned by the vast majority of early childhood professionals—or ignored in the publications and research of such groups as the National Association for the Education of Young Children. Objections to television viewing in institutional settings reflect the ambiguous status of these childcare spaces as intermediaries between the home (where television viewing is usually frequent, acceptable) and the school (where explicit learning, rather than merely childcare, is supposed to be taking place). The quantity of television viewing in daycare settings is frequently used to evaluate the quality of centres. If too much television viewing is done, middle-class parents often complain to teachers or centre directors. Some daycare centres claim superiority over home-based daycare, based on their restrictions of TV watching. For this reason, the corporate daycare chain Kindercare, which now claims 1 per cent of the entire market, has written policies limiting the amount of time videos may be viewed at school, and specifying what types of video may be watched (such as those with general exhibition [G] or parental guidance [PG] certification only).

Those who care for and teach young children interpret and enact their jobs differently, with the most highly paid and educated women emphasizing their role as teachers, and the lowest paid emphasizing the type of care associated with mothers: such as providing affection, nutrition, and training in personal hygiene (Wrigley 1990: 304–5). Upper-middle-class pre-schools also seem to accomplish some of the important work of inculcating tastes that was formerly reserved for the home; while schools with working-class students emphasize teaching of letters and numbers.

Although I will discuss a male teacher in this chapter, childcare and teaching remain overwhelmingly pink-collar (i.e. jobs traditionally filled by women, such as clerical work and nursing) work. Sociologist Randall Collins has pointed out that many occupations performed by large numbers of women are inadequately described in conventional class categories, since the pay may be low but the prestige relatively high. Status plays as large a role as wages in pink-collar work such as arts administration or interior design or teaching. The location of these occupations in the cultural sphere is extremely important and such female labour may be inadequately compensated in monetary terms, but play a vital role in the production of status culture:

> Women live subjectively—and, in terms of their successes, objectively as well—much more in the realm of status than in the realm of class. This might make it seem that women are living mainly in a realm of illusion, a cultural fluff floating over the hard material basis of our society. But the capitalist economy of the twentieth century has increasingly derived its dynamism from the permeation of status symbolism into the material objects of everyday consumption . . . The activities of women, in both the production and consumption of status culture, may well constitute the feature that keeps modern capitalism alive. (Collins 1992: 229)

My research suggests that status production is an important part of the work of professional pre-school teachers—as it is with many pink-collar occupations—and the aspect of their work that may be most valued by the women themselves.

Within these parameters of teaching and childcare and the culture of the school or centre, then, complex attitudes and beliefs about media effects evolve—with those most interested in status production and who deem their job to be the most professional also the most critical of the media. In what follows, I compare four case studies of three women and a male caretaker, to outline the range of lay theories of media effects, their derivation in other aspects of an individual's 'media biography', and their correlation with other kinds of ideology.

A Montessori Pre-school

Sara Kitses has taught at the suburban Montessori school for twenty-three years. Her half-day classroom combines children between the ages of three and six. She has a total of fifty students in her morning and afternoon classes. Sara describes her students as representing 'very little economic diversity', mostly white, with a number of 'oriental children', often faculty children. Tuition is about the highest in the community, nearly $400 per month for less than four hours a day. Sara is the highest paid teacher in the community. The Montessori school is attended largely by the children of attorneys, physicians, and university professors. Sara is about 50 years of age, and holds a bachelor's degree in addition to specialized Montessori training.

More than any other teacher in the study, Sara denied having any knowledge of popular children's programming (she asked 'Who's Barney?' during the interview). In her long experience in the classroom she has picked up a passing acquaintance with some of the names of programmes and characters she herself has never seen—usually to marshal forces to ban them from the classroom. She rigorously excludes videotapes from the classroom, even banning the widely accepted Disney feature films, *Disney Sing-Along* videos, and PBS programmes. Such videos are so widely accepted that they were included in the libraries of every other centre I visited. Disney films have achieved flabbergasting levels of market penetration—as well as acceptance by adults as unobjectionable material. During the interviews, teachers were asked to check off tapes from a list of fifty currently popular videos with which they were familiar or had shown at school. The only video from the list that Sara had ever shown was an animated version of *Dr Seuss*.

Sara requires parents to attend a meeting before the school year begins (unlike many centres, where children come and go from month to month, Sara's school enrols children only in the autumn). Sara advises parents that the best thing to do with television is keep it off, but the least they must do is monitor what the child is watching very carefully and remain with the child during viewing. Sara commands considerable respect from the parents in her school, and openly claims the status of an expert with regard to the children —one who knows more than the parent. She has a great deal invested in her identity as a professional teacher. The Montessori school is typical of that segment of the childcare market which promises preparation for entry into the competitive world of the grade school. Sara's status depends in part on the degree to which parents grant her the power to evaluate their children's social and academic skills. Television threatens both the children's cognitive development and her own professional standing—if children can learn from videos, and be entertained through their childhood years, why would they need a special theory of education? Sara's mind was made up a long time ago to the view that television is a very negative influence on everyone—for more than a decade of her married life and throughout her children's early years she had no television set.

Sara's disdain for television dictates her selection of classroom media. She prefers showing 16 mm films and film strips to the students (she estimates she uses film three or four times more often than video), nearly all of them adaptations from children's literature. Videotapes are restricted to programmes such as *National Geographic* specials. While Sara spoke about exhaustion in her job, she does not consider media screenings to be appropriate for teachers to use as a break to do other chores in the classroom (she has a large number of assistant teachers in her class). She monitors the screenings vigilantly, and operates the 16 mm projector herself. While Sara retreats behind the projector, her assistant is in charge of supervising the children during the screenings.

Children and parents bring videotapes to school, and occasionally Sara will show the tape if it is deemed by her to be educational, if she has prescreened it, or if the parent is someone she trusts. All her recent examples of videos she allowed were non-fiction: a computer-animated film with an exclusively musical soundtrack by Philip Glass, distributed by the upmarket Nature Company; *Road Construction Ahead*, an independent video for children (featuring no words, just music). Children can bring objects from home every day to set on a viewing table. The rule is that whatever is brought to school must be educational, a concept that, according to Sara, the children and parents have no trouble grasping.

Sara's complaints about television and children cluster around issues of fantasy and passivity. Primary is the objection that television comes from adults, it does not originate with the child:

> I really believe there's a lot in children, and if you give them a proper environment, they will act upon it and will be constantly learning. They're very eager to learn. And if what you give them is a lot of television and passive stuff, or it's all coming at them, they have to make no decisions whatsoever except to turn it off or on. I don't think that's . . . [what] we should be advocating.

Television, Sara believes, is commercial and makes children want to buy things they don't need; Sara is interested in environmentalism and reducing material wants. She thinks that television is frightening for children, and introduces them to material inappropriate for their age. She even dislikes *Sesame Street* for its 'silliness' and its presumption that children's attention spans are very short. In addition, television is simplistic in its advocacy of violent solutions to problems, and Sara finds the sex-role stereotypes and focus on appearance and decoration offensive in the media targeted at girls. Most importantly, television isn't real.

Sara believes children have a very tenuous hold on reality at this age, and the proper role of teachers is to guide them towards reality—thus her heavy emphasis on non-fiction materials. Television pushes children into fantasy and leaves them 'perpetually confused'. Television stories, unlike some more naturally occurring kinds of role play, do not equate with creative play:

> Certainly, make-believe and role-playing are wonderful and important things. But I don't consider that a fantasy that's being perpetuated by adults. That's coming from within the child, and fantasy coming from within the child is a good thing, but when they are playing games, like Ninja Turtles, they start playing Ninja Turtles on the playground, that's not creative play. It's something that's been given them, so I make a distinction between those two things.

Sara backs up her feelings about television with some stiff rules: no talk about television on the playground, no television play, conversations with her about television programmes are discouraged, clothing with media characters are not allowed because they distract children (parents are informed of this on the first day). Sara spoke of one boy who was obsessed with what she believed was Nintendo, and would talk at great length to the other children about the story and the characters. Sara was annoyed: 'It was very boring. The other kids didn't like it.' When the boy turned to talking to the teachers, Sara became increasingly forceful in limiting such talk:

> I had decided that I had a limit to listening to him. So when he would come to school, I'd let him talk and then I'd say, 'Let's focus on the environment and what you might . . .' I didn't tell him not to talk about it, I just redirected his attention. 'What would you like to choose today? Can I help you make a choice?' Trying to focus him in the classroom. Finally it came to the point where we had to say to him after that initial time, 'While you're at school, we want you to just be talking about school things', because he was throwing the other kids off.

Reading between the lines, I would suspect that the other students' reactions to such television talk may not have been solely boredom, but that some amount of imitation of this unwanted behaviour led Sara to lay down the law.

Curiously, these rules apply much more leniently to girls, who tend to play in groups separate from boys. Sara knows that the girls play games in secret based on *Aladdin* or *Beauty and the Beast* while they are out on the playground. But the girls only play these games out of her hearing, and, because it does not create disruptions, she does not intervene. Also, Sara does not enforce the no-characters rules with the girls' clothing, allowing Jasmine and others to make their way into the classroom on shoes, socks, and t-shirts. She does enforce a ban on jewellery, and encourages parents to dress girls in slacks and sensible shoes.

Sara explained that there is a costume area in the classroom, stocked with Greek, Indian, and Native American costumes that she supplied herself. When the children enact a play, the teacher generates the story, acts as the narrator, and assigns the parts to the children: *Three Little Pigs* or *The Billy Goats Gruff* or *The Little Engine That Could* are examples of stories she might use. I was wondering why Sara saw no contradiction between these guided

Child's drawing of action hero: creative fantasy?

forms of dramatic play and her position on outside influences. Does an interest in Greece come from within the child, for example, or is it Sara's? Is acting out parts in a pre-scripted story any more creative than acting out parts in a television play? When I first introduced myself to Sara, I explained that I believed TV effects were often misunderstood and overestimated, so that I could discuss what I perceived to be contradictory positions openly with teachers during the follow-up interview. In this exchange I was more openly and directly challenging than in any of the other interviews, in part because I disagreed with and had the least rapport with Sara, and in part because she seemed eager to stake an intellectual defence of her position.

SARA: I like to see them doing role-playing kinds of things that might put them in an adult model. House kinds of things or positive adult kinds of things that where they're enacting something good instead of something horrible. I don't like it when they get involved in fantasy that's created by the media, because it doesn't go anywhere, and it's not coming, again, from within the child. They're re-enacting something they've seen. Part of that's inevitable, part of it's how they learn. But—

ELLEN: How could you explain the difference, then, if they're acting out something that's from a book, like you said that you do plays from books?

SARA: That's a fairly structured activity when we do these plays. You take on this role and you are that particular role in the play. You're supposed to follow the storyline. We do that at two levels. One is just informally, in the circle, and the other is this play that these older kids put on every year. And they understand that quite well.

ELLEN: So you see this in a different category because that's a structured—

SARA: Exactly, the teacher is guiding the children into that, whereas the other is out on the playground.

Sara is very clear that her role in the classroom is to provide authority and to keep the children under control. Television—as a video shown in class or as references in play or on a t-shirt—symbolizes a loss of control for Sara. Not only is television a world she is almost completely unfamiliar with, but—without viewing it—she has an overwhelming feeling that it undermines everything that her own classroom stands for. Sara's complaints are not the more common ones about rowdy behaviour or aggression, although she mentioned these in passing. Instead, they are phrased in the language of child psychology: TV delays the achievement of the developmental stage when children can discern the difference between reality and fantasy. But her goals are embedded in the dispositions of the class to which she and her students belong. By emphasizing creativity as a spontaneously produced result of children's play, she legitimates her role as an educator in aesthetics as well as behaviour.

In Sara's scheme of value, books are basically good and television is basically bad; film falls somewhere in between. Unlike most of the other teachers, however, for whom any reading is viewed as positive, Sara is unwilling to accept just any books. She discourages the purchase of 'grocery-store books' such as *The Berenstain Bears*, as well as any media-related books, such as Disney stories. While her principles emphasize the generation of fantasy from within the child, she approves of the enactment—on the playground or in the classroom—only of stories from canonical children's literature: these forms of imitation are not recognized as imitation *per se*. Sara prefers older classics. *The Little Engine That Could*, which she reads to the children and has them enact as a play, could be seen as a consumerist fantasy—the story concerns the obstacles to the delivery of a trainload of toys and goodies for children at Christmas. I suspect that the age of the book, and its lack of adaptation into a well-known film or television form, work to exempt it from criticism or scrutiny. Non-fiction is preferred to fiction in the choice of videos, and for computer software. The children are encouraged to write 'research papers' on topics of interest to them: this, I would suggest, is the highest activity on Sara's scale of value, combining as it does writing skills, an ordered display of encyclopaedic knowledge, and the rehearsal of forms of work that will certainly be required in grade school.

If television is the school's bad screen, to be used sparingly and purposefully, computers are its good screens, and access is open throughout the school day, limited only by the number of terminals available. Sara spoke with pride about running 'ahead of the pack' because her school budget allocates money for computers, which she referred to as 'an absolutely critical moment of information right now for young children'. The classroom has one computer with a CD-ROM, and another computer set up with WordPerfect and Logo, software that teaches computer programming. A sign-up sheet and two chairs are placed at each computer, but the children are allowed to work alone if they wish. Sara is 'flabbergasted' by the speed at which children learn to operate the mouse, and she uses computers heavily for teaching alphabet and reading skills.

Sara says the computers are fun for her because they represent something new after so many years of teaching. She compares CD-ROM to video this way:

> They're interactive. They have to make some choices as to what they want to do. It's much more informational than videos, so they're learning a lot. And there are levels of difficulty, so I find that the 5- and 6-year-old is going to a higher level of difficulty, so he can read and . . . They'll choose things that are challenging to them and interact with them. They're much more popular with boys than girls, but it might be because . . . I don't have any programs on the CD-ROMs yet that are other than mammals and the San Diego Zoo . . . [and] the dinosaur program—the boys tend to dominate it. I think it's better than videos because it's much more interactive, but it can't be their entire life.

It is not surprising then, that virtually no television viewing was noted on Sara's media diary, although she had shown a non-fiction tape, *Road Construction Ahead*, on two occasions that week in her classroom. Sara had warned me at the first interview:

> You're going to find me, I hate television. I really want to devote my time outside of school to artistic endeavors. I'm a photographer, and I've got this incredibly strong feeling that if I don't use my time well, I'll never use it. I'll just be like everybody else, watching all this stuff, and I read a lot. But that's a choice. I choose what book, I don't turn on the television and make somebody else make those choices. So I am kind of extreme.

Sara's resistance to completing the media diary was reflected in her failure to bring it with her to the follow-up interview. She did not begin to keep the diary until several days after our interview. In it, she records listening daily to NPR, both *Morning Edition* and *All Things Considered*. Her reading that week was Carl Jung's *Memories, Dreams, Reflections*. She watched two rented videos that week, both of them costume dramas, art films: *The Bostonians* and *Howard's End*. When pressed about any passive viewing she might have done Sara explained that her daughters watched *Oprah!* several afternoons, but she remained apart in the kitchen with her radio.

Despite Sara's harsh judgement of contemporary children's television, she spoke fondly of her television viewing as a child. Her parents paid little attention to what she watched. Television was a novelty then, and her family was early in getting a set. She never watched cartoons, but she remembers *Ozzie and Harriet*, *My Friend Flicka*, *Lassie*, *The Ed Sullivan Show*, *Superman*, and *The Jack Benny Show*. *Lassie* was the show that she felt a strong affection for and she believed television had a positive effect on her:

> Actually I think that it had a positive effect, because I was seeing children in other families and parents that were relating in a very idealistic way. Things were wonderful, and everyone was looking up to a glorious future, so I feel good about the television that I watched, especially *Lassie* because I loved animals so much. This dog did all these wonderful things in relationship to this boy, and his relationship to the family.

Sara's image of television during its golden age is overwhelmingly positive, and, although she has so little experience viewing television over the past twenty years, she compares it negatively to contemporary television: 'They addressed heroic kinds of things, solving problems, moral values, ethical values. I think a lot of that has gone.'

Sara's belief system about children and the media typifies that of those with the most training and investment in the professionalization of childcare. She subscribes to a developmental view that emphasizes cognitive deficits, and thus children's incomprehension of television and films (Anderson 1983: 395–400). She constantly interprets children's popular media through the 'frame' of developmental psychology. Sara's unfamiliarity with TV and her refusal to listen to the children's talk about it, however, make for a very limited basis on which to formulate her judgements. In fact, Sara herself is somewhat incompetent as a TV viewer, and she uses this studied ignorance about television to signal her erudition—she fears being 'just like everybody else'. By completely discrediting knowledge about television in the classroom, and limiting media use to those forms—such as 16 mm films—that are more difficult to operate, Sara also maintains her authority over the children.

Hodge and Tripp (1986) construe television as a barrier between teachers and students; middle-class teachers understandably might wish to bar television from the classroom in part because they know less about it than their students do:

> Not only are they untrained to deal with it, but it so often challenges their own knowledge and experience of life and understanding in general. In comparison with most teachers, the average pupil watches a great deal more television and quite different television programmes. Whereas 20 years ago teachers could, in a very real way, safely assume that their pupils knew a great deal less about everything than they did, today the children often know a great deal more than their teachers, albeit about things such as Batman, Spiderman, Superheroes. (1986: 170)

Hodge and Tripp found that while teachers do not overtly punish pupils for bringing their television experience in the class, signals are nevertheless clear that television is not a legitimate topic for discussion at school.

Running underneath the discourse about children's incompetent distinctions between reality and fantasy, and the fragility of their comprehension of narrative, is Sara's belief in the need for adults to shelter children from all that is too intensely emotional, too dramatic, or too peer-oriented because of their moral deficiency, their easy corruptibility. Thus, an essentially romantic view of the innocence of children is masked by the discourse of Piagetian cognitive development.

Reinforcing these notions of deleterious media effects is an equally strong belief in the need for explicit teaching in the area of aesthetics. Systems of distinction extend to clothing (should not be too feminine, should not be too representational—not purchased in discount stores), to books (should not be bought at the grocery store or the mass market). Sara is heavily interested in both what is acquired and how it is acquired—a distinction that increases in importance with status. She confines herself to a very narrow range of media forms, those most closely allied with books. She prefers 16 mm film and CD-ROM as modes of delivery over videos, despite the greater ease and affordability of video. As Bourdieu points out, the more difficult the means of acquisition, the more distinguished the cultural goods. Sara risks the appearance of authoritarianism (with her dress code), arbitrariness (with her censorship rules), and severity (with her banning of toys from the classroom). But if her job is viewed as one of status production and the inculcation of taste, the strictness of her rules and her lack of involvement with nurturing behaviour begin to make sense.

Girls and boys mount different forms of rebellion, different susceptibilities to vulgarity. The girls gravitate towards things that violate the aesthetic code (Disney characters) and the ideology of gender neutrality (frilly dresses). The boys tend to violate the decorum of the classroom by moving in too close to others, being too loud, dominating conversation. The boys' attachment to the media is pathologized because boys more often require direct intervention by the teachers to bring them under control. The genres boys gravitate towards as media fans—science fiction, superheroes, action-adventure—are associated with the derring-do masculinity that is especially disapproved of. These boys are encouraged to develop intellectual, not physical, abilities. Girls simply receive less attention altogether—Sara is disappointed by their lack of curiosity about the computer, for example, but she also finds that the girls conform to her standards as good students more readily.

Within the upper-middle-class milieu of the Montessori school, enforcing the ban on television should not be viewed as a marginal part of Sara's job or a personal idiosyncrasy. The TV ban is—as Sara herself believes it to be—a crucial element of her work. Its justification in terms of theories of childhood development masks a more class-conscious motivation, that of distinguishing these children and their education from that of the common mass. As Randall Collins describes this type of work: 'The higher classes . . . observing the cultural style of the classes below them, engage in reflexive role distancing, once again re-establishing their superiority to those who have a less sophisticated view of cultural symbols' (1992: 217). Sara's vigilance in enforcing these norms may pit her against some parents at the school (who are left with many more hours of the day in which to occupy the child while keeping them away from undesirable media, and who generally fail to hold to Sara's high standards), yet on the whole they acquiesce for the sake of securing a successful future in the education system. Television is a key symbol in the very explicit taste- and class-distancing that the school provides.

'I'm Not the Morality Police'

Glenn McClintock teaches at a 'developmental learning centre' down the road from the Montessori school. The clientele consists of professionals—university professors, doctors, lawyers, architects. The school serves the same economic stratum as the Montessori school, but serves more dual-career families because it is possible for children to remain in care for a full workday, and the school accepts children from infancy through kindergarten. The school administration has tried to remedy the economic homogeneity of the student population by offering a handful of scholarships each year to low-income families. There is a counter-culture, an alternative feel to the school: teachers dress casually in jeans and t-shirts, children's artwork covers the walls of the permanent facility.

Glenn is white, single, 30 years old, and a college graduate who has taught at the school for ten years—a very unusual achievement for a man in this field. At the time of the interview, Glenn had twenty students in his classroom of 4-year-olds—but he had had experience working with all the different age groups at the school. He seemed well prepared—even eager—for the interview; he knew what sorts of question to expect because he had been recommended to me by another teacher in the school. I also served on the board of the school (as an outside academic expert)—having been invited by the director after starting the interviews—so Glenn knew about my research from the school newsletter, and might have read my book, which the director kept in the school's library (Seiter 1993). My position on the board also placed me in a position of heightened authority in the interviews, since I was well acquainted with Glenn's boss.

Glenn's school had a very large playground area, and students were encouraged to play outdoors even in cold weather. Art and nature activities were given a high priority in the curriculum. The school placed a premium on receiving feedback from parents about the school's activities. One consequence was that parents quickly objected when teachers showed too many videos in the classroom. Glenn explained that the primary feeling was that watching videos was something they could do at home—without paying tuition for it. Nevertheless, the school kept a television monitor and a VCR attached to a cart that could be wheeled into a classroom or moved to a general purpose area (here called 'center court'—the multipurpose room). Glenn was going on a rule of thumb (one that had been articulated by the director) that videos should only be used every three weeks—although his video diary indicated more frequent screenings than that. Because every classroom at the centre has two teachers, one routinely reads a book to the children, while the other sets out lunch and cleans away the dishes—a luxury made possible by the school's larger staff numbers.

Glenn's school houses a library of approved videos, which include *National Geographic* specials, and *Dr Seuss*—Glenn's favourite. As with the Montessori media offerings, a strong preference is given to videos that derive from widely known published materials—especially classic children's books. The public library's bookmobile visited the school weekly and offered a variety of videotapes that teachers could check out. Disney videos were brought in from home by the children, and sometimes they had movie parties with popcorn. During the week that Glenn kept a media diary, the teachers had brought in a tape of the Olympics to show the children who were interested in sports.

The developmental centre's rules for viewing videos were less strict than at the Montessori school, but Glenn's critique of popular media was just as strong. He was caught between a permissive anti-authoritarian model of education and a belief in the harmful effects of the media in three primary areas, reiterated throughout our interview: (1) the promotion of war and guns, linked to the theme that might makes right; (2) the cultivation of female narcissism and an overvaluation of physical beauty; and (3) stimulating children's out-of-control consumer desires. There are strands of Sara's critique of children's media (especially about consumerism and a negative immersion in fantasy), but Glenn's frame is a moral one, more so than the developmental psychology frame Sara employs. What they share, however, is a lay theory of strong media effects.

The condemnation of superhero cartoons used to promote PBS

Glenn was deeply critical of superhero cartoons, video games, and the general category of masculine action-adventure drama. His criticism was based on an ideological reading—one might call it pacifist—of the shows, in their promotion of a simplistic moral worldview and violent solutions to conflicts:

> I don't know, some of the shows really bother me because I think they're very black and white and they teach this sort of 'might is right' and it's OK to hurt somebody as long as you're the good guy . . . And it goes along the same message that if a person's bad, he's completely bad . . . they always want to take over the world. They're always completely reprehensible and so, the Turtles feel like they're justified in kicking in the face or doing what they have to do. I don't think that's a particularly good message for children, especially young children because they have a hard time anyway discriminating between fantasy and reality. I think they really identify with these characters because they can kind of do what they want. And I think that's real attractive to a little kid because

> these characters are never afraid and they're very powerful. And these are things that young children are not.

Glenn's lay theory involves both a sense of children directly learning violent forms of conflict resolution—learning to fight instead of using words to settle disputes, as he teaches—but also a sense of indignation about the media's corruption in preying upon children's feelings of vulnerability. (There are echoes of this in Sara's understanding of childhood, as well.) This indignation reaches its strongest expression in Glenn's feelings about guns—he immediately puts a stop to gun play when he sees it and he is very outspoken with the children about his feeling that guns are bad. His model here is that if children play with toy guns they will be likely to use real ones, that good parenting involves setting clear limits based on ideals such as non-violence, and that parents are often too busy or too wealthy to set such limits:

> 'Cause I believe that a gun isn't a toy. That to teach them that message is really harmful because you know pointing your gun at someone even in a kidding way is just not, not OK in my book. 'Cause I think you know if they ever did get a hold of a real gun they'd want to do the same thing and it could be horrible. Horrible consequences . . . 'Cause the other thing I mean that really scares me is the sophistication of some of these toys. It's not just like a cap gun. They have these things that look like, that you can buy that look like an M16 or a real machine gun. That really scares me that that's available or someone would think that is appropriate for a little kid. I think, you know, I still think that the primary responsibility falls on the parents. And they have to say, you know, if you're going to cry or maybe the next door neighbor has this thing but that's, that's too bad. That's what my father did. He served in the Second World War and he obviously used guns, but he said, 'Forget it, you know, you're not going to get it. That's too bad.' I think the peer pressure is greater today because it's, they've really, they've targeted, they are taken, this is a very marketable age group. You've got a lot of parents, you know, middle class that they have money and . . . the kids. You know it's not just like the teenagers having more where, you know, [they] buy the latest clothes. It's the little kids. 'I want the Cabbage Patch doll' or 'I want the Aladdin figure'. These are not cheap. Forty or fifty dollars if they want to have it, and it's hard to, I guess it's hard for some people to say no when they go into a toy store. You know, rather than deal with a tantrum they give in sometimes.

Glenn's critique of media violence blends easily into his sense of children's desires for consumer goods spiralling out of control. A regular reader of the *New York Times*, Glenn also referred repeatedly during the interview to the ways in which Desert Storm was perceived through media coverage as similar to a video game or a television show, and how this had desensitized people to violence. Particularly appalling to Glenn is the idea that video games might have been 'modeled after Desert Storm'.

Glenn recounted recent play scenario in his classroom with sword play and with wrestling on a rooftop (the final scene from *Beauty and the Beast*), which he allowed to proceed so long as no one was getting hurt. Glenn had been recommended to me by an earlier interview subject as someone who had a different—more tolerant—policy about rowdy play in the classroom. Glenn reported that children enjoyed his presence in the school because he engaged in more physically active forms of play—chasing, throwing them in the air, wrestling—than other teachers. He even mentioned that he thinks some female teachers worry too much about children getting hurt. If a gun appears in the play scenario, however, even as a symbol (a piece of bread, a Lego turned into a gun)—he intervenes. One part of the lesson he is teaching, then, is that symbols are extremely important, even in play, and that fantasies expressed in play scenarios are to be controlled—even more significant than the more mundane concerns of personal safety. This is quite similar to Sara's point of view on the necessity of limiting fantasy and bringing children around to reality.

The media brainwash children with advertising, in Glenn's view. He is dismayed to see 5-year-old children pretending to be Shaquille O'Neal when they play basketball; he laments the prevalence of Disney and the ways in which children automatically prefer Disney films to other videos he chooses, because the characters are so familiar to them; he notes a constant stream of new toys and toy requests from the children, and finds contemporary toy stores completely overstimulating to children.

> Yeah, I think the big difference that I've noticed just working with kids —I've worked with kids for 10 years—is that I think there's so much, there's a lot of emphasis placed on marketing, marketing toys through shows. It's almost as if, you know, Madison Avenue sort of decides these things, you know, develops a product and then you develop a show as an afterthought to some product . . . I mean, certainly when I was a kid there were things that you could buy that are on TV shows, but I think it's just so much more prevalent today.

While Glenn has no problem drawing a firm line with the rule on toy guns, he is somewhat more unsure about expressing another one of his convictions—that children's consumer culture places too much emphasis on physical appearance and female beauty. It distresses him to hear girls claiming 'I'm the prettiest' in his class during play scenarios, and he regularly steps in to squelch such expressions of power:

> GLENN: They'll say 'Oh you don't get to be, you don't get to play in this game' 'cause this is kind of the way it goes. I think generally girls tend to, they use their words to hurt, whereas boys are the more physical, like 'I'm gonna push you down'. But with the girls it's more like 'You can't play. I don't want you to be a part of it. So everybody play except you.' And you are kind of left out.
>
> ELLEN: So what would you do to, how would you handle that, like if they were saying 'I'm the prettiest'. What would you say to them?
>
> GLENN: Well, that's the thing that's kind of hard. I would probably say that I don't, I just try to emphasize that it doesn't really make a difference. That those are things that don't really matter that much. That we're all friends here. And we don't, we try not to compete in that way.

Glenn's feelings on the topic of female narcissism has been much influenced by one of his fellow teachers. Curiously, he had more to say about this topic than any of the women we interviewed, and he clearly articulated a critique of artificial beauty that seemed one part counter-culture and one part feminism. The teacher he so much admired was herself 'stunningly beautiful', but her beauty was effortless:

> One of the problems I have is, is with the female characters are known as, are always like really really beautiful and it seems like that's such a big part of their character, you know, like finding a prince, like in *Aladdin* finding the perfect man, and it's not really, I don't [know] if that's a really good message to teach a little girl or not. Also, I notice some of the extremes in some of these movies that if you're beautiful you also tend to be really really good . . . 'Cause I just see that as a problem in society generally that so much attention is placed on a woman's looks instead of, is she competent or what she can do. These movies feed into that. And the Barbies and the Ariel and Jasmine, I mean they all feed into that, you know, looking good is too important.

Glenn has some uncertainty about how far he himself can go in voicing this critique without hurting the little girls' feelings: 'I really don't know how far I should go as far as saying I don't like these, 'cause . . . They're so sensitive about a lot of things. I don't really want to say I don't like that 'cause it might be kind of like "I don't like you".' This is not a conflict he feels with the boys when he tells them that he does not like guns.

Glenn faces conflicts with parents over how far to take his critique of the media. While his views implied a theory of strongly detrimental media effects, he was unwilling to take this—as is done at the Montessori school—into any attempt to press his views onto the parents or their home life. He refused to take on the traditionally female role of parental adviser or social worker. As he put it: 'You know I don't have any jurisdiction over what people do outside of here, nor do I want to either.' He mentioned a number of instances where parents were embarrassed to reveal that their child played with action figures at home, or was familiar with superhero characters, as in the following story about a mother's embarrassment when her son's Batman figure accidentally wound up at school:

> GLENN: She said, 'Well I didn't want you to know that he had one.'
> ELLEN: So what did you say?
> GLENN: I said I don't really care. I'm not the morality police [*laughs*].

Glenn mentioned later in the interview, however, that he believes it is much easier for children if the same rules apply at home and at school, but that he found it distasteful to try to exert any authority or influence over the parents or their home lives.

Glenn's own choices of leisure activities favour middle-class, wholesome, respectable forms of masculine endeavour reflecting a sense of doing, rather than just watching, as an important principle of the worthwhile. He consistently rejected the action-adventure genres or casual TV watching he disdains in his students' lives, and prefers playing basketball, jogging, and appreciating nature. His favourite activity on the job is teaching the children about the environment: planting seeds, going for nature walks, taking them to the zoo, and teaching them about whales. It is in this regard that he feels his work may be most significant, too, in 'planting the seeds' of environmentalism at an early age.

Glenn's media diary revealed a highly respectable assortment of activities: listening to classical music and jazz; seeing *Schindler's List* and *Philadelphia* at the movies; watching local college basketball and the Olympics on television, reading the *New York Times* and an F. Scott Fitzgerald novel at bedtime. Glenn's childhood memories of television are not as warm as many others we interviewed. His favourite show was *Rocky and Bullwinkle*. He watched 'mostly superhero cartoons' such as *Super Spiderman* and *Batman and Robin*, *Captain Kangaroo*, and sitcoms such as the *Addams Family* and *Gilligan's Island*, noting apologetically that there wasn't much (worthwhile) available in those days before *Sesame Street* or *Mr Rogers*. He spoke most enthusiastically about playing outdoors, and playing sports.

Glenn was conscious of his unusual position as a male pre-school teacher and felt that he was appreciated for it by parents and children, although he also faced some discrimination—a sense that he wasn't up to the job of caring for small children. His work has led him to appreciate fully how difficult the job of caring for children can be. He said that he now believes it is much easier for a parent to go to work outside the home each day than to stay home and care for children. His views on child-rearing, however, favour a conventional arrangement with a single wage-earner and a full-time homemaker. His own memories of a middle-class childhood with a mother at home lead him to speculate on the toll on children of being in care all day. Like Sara, whose mother was home all the time caring for her as a child, Glenn has a sense of having had an easier childhood because he was not in any institutional settings:

> Yeah, in an ideal world there'd be one person, one'd work and the other person could stay at home and be with the children . . . it's not easy for [the children] to be here 8 hours a day . . . When mom is gone, she's gone. And there's no coming back until 5.00 you know. That's, that's a, that's kind of a, a large dose of reality for a little kid to face, and, you know, sometimes they just get upset and want to be with their parents and then there's not really a whole lot you can do about that 'cause you're not their mother or father. That's tough, but they deal with that in the best way as they can and you try to help them deal with that in the best way you can.

Gloria's Family Daycare

Along with her 74-year-old mother, Gloria Williams owns and teaches in a centre called Gloria's Place, which operates in her own home in an older, integrated neighbourhood close to downtown. This centre is a federally funded provider that offers twenty-four-hour care for children from infancy upwards. In the interviews, Gloria relates with pride that she is the only licensed black childcare-giver in town. Her influence in the community extends well past the children into the lives of the parents, most of whom are divorced. All of her mothers and fathers work for pay outside the home—unlike the Montessori school, where the half-day programme necessitates having one parent at home (meaning, usually, that the father is receiving something that can serve as a 'family wage').

A 43-year-old woman from New York with a degree in computer programming, Gloria switched from her office job to childcare at the age of 25. She is not college-educated in early childhood education, although she is well read on the subject and receives a lot of training in the form of workshops as a daycare provider eligible for federal subsidies under Title XX. Gloria's mother works with her and has taught her a great deal about child-rearing. Recalling her own childhood, Gloria mentioned that there always seemed to be children for whom her mother cared around the house. Gloria charges a sliding fee: 'I can't see a person making $200 a week and then you're taking over half of that in daycare, you know, I can't, so I have a scale where I go with the person's income.'

Gloria conceives of her work very broadly, combining the role of foster mother and teacher. She strives to prepare children for school by teaching them letters and numbers and encouraging them to be independent from adults. She emphasizes giving the children experiences such as trips to the library that their parents don't have time for. Gloria's interview took place during Black History Month and that week the library books she read to the children clustered around that theme: *The Black Women's Poetry Book*, *The Drinking Gourd*, and biographies of Sojourner Truth and Dr Martin Luther King. But she also emphasizes the work of providing nutritious food, a safe environment, and affection. She was more frank than any other teacher we spoke to about how trying this kind of work can be:

> Some days . . . I mean, if they had a bad day at home, they're going to try to carry it on through here. [*Mimicking the children's voices*] Don't look at me. Don't touch me. Leave me alone. He's looking cross-eyed at me. What's he staring at? He touched me. He's got my Crayola. I was playing with that block. I don't want this. I'm not hungry. I don't like this kind of juice. Why can't we have chocolate milk? Stuff like that. Things like that [make] you want to just put their coats on and march them right back out. So some days it's not worth getting up, but you know you have to.

Gloria considers the local wages to be especially low and says that she made twice as much, with all the advantages of a regular schedule (such as weekends, evenings and vacations free) in her old job. Typical of family childcare providers, Gloria is threatened by many of the hazards of this type of work: burnout, emotional attachment to children she lacks the authority to protect, frustrations stemming from low pay and inadequate funds to care properly for children (Nelson 1990). But it is clear that a primary incentive is her dedication to the children she cares for, both as a foster mother and as a daycare operator.

Gloria's approach to television viewing was keenly interested and unapologetic. She shows tapes frequently and lets children choose among television broadcast and cable channels. (Home-based daycare is much more likely to have cable hook-ups than public daycare centres, thus greatly increasing the options for television viewing.) Of all the teachers interviewed,

Gloria was the biggest fan of *Sesame Street* (which she much prefers to *Barney and Friends*), and her children watch the PBS line-up (as well, they visit the local PBS station each year to appear on air during fund drives). They also watch commercial television and choose from her large selection of purchased videotapes (including every Disney movie ever released).

Throughout the interview, Gloria chatted freely about a very wide range of television programmes and films; she was full of information and opinions about them. She had strong preferences about children's TV: for *He-Man* over *Power Rangers*; *Captain Planet* over World Wrestling Federation characters. She does not impose her preferences, however, when the children select programmes to watch. Television is used in the mornings, when children are arriving; at lunch time to 'quieten them down'; and at pick-up time around 5.00 p.m. Sometimes they have special character days, when they dress in special clothes or colours and enact stories based on their favourite characters. There are no restrictions on clothing or on toys, and Gloria is familiar with the wide range of items—shoelaces to backpacks—with licensed characters, as well as the latest toys and what is available at stores in town. She lets children bring anything to the centre with the exception of toy guns and knives.

In discussing children and television, Gloria refuted the argument against fantasy as a developmentally dangerous activity. Gloria herself is familiar with a broad range of film and television genres and gave the most informed and sophisticated readings of popular culture—those closest to the opinions of media studies academics, in fact, of all the teachers we interviewed. In striking contrast to Sara's concerns that children become lost in television fantasy, Gloria described her students as easily able to understand when something is 'nothing but make believe'.

> Well, you've got to go on the cartoon thing that they know it's make-believe in the first place. And I think they're cutting the kids really short if they don't think kids have it in their brain that this is make-believe, that you cannot do it. It's like *The Three Stooges*. We knew we couldn't do those things and still be alive with a normal brain. That's the way it is with the cartoons. And the parents should know that they do have some imagination. And that it's nothing but make-believe. There are some violent cartoons, but you know the person (cartoon character) always gets up and goes on about its business or you see it (alive and unhurt) in the next section. So I think the child knows that it's make-believe.

Gloria reported that children talked about television or movies every morning when they come to her place, and frequently described to her the plots of entire movies they have seen over the weekends. Gloria perceives her children to be much more sophisticated about their viewing and understanding of genre rules than Sara perceives the children at the Montessori school. One reason may be the relaxed atmosphere at the centre and Gloria's openness to listening to conversation about television.

> ELLEN: Do you find the kids are pretty good at telling the stories of things they've seen?
>
> GLORIA: Yeah. A lot of them are. Some of them just roll two stories into one. But I usually let it go like that if they're telling the story,

> they're telling it their way. If I'm reading a book I'll straighten out the story. But when it's their time to tell the story I let it go. But it depends on when they want to talk. It's just when they want to, it's right there. But if you ask them on command, forget it. They're not going to do anything. You know . . . or you can creep up on them and hear them telling the stories. But if you ask them to get up front of their own peers to tell the same story—forget it.

Gloria's careful attention to children's play scenarios has also influenced her opinion of television's promotion of violence in children's play. She seemed to have a more accurate, more vivid memory of play—and of movies and television—than many of the other teachers we interviewed. Her childhood as one of seven children—with four brothers—may also have promoted a much higher tolerance for physically active play than other teachers we interviewed. She thinks of rowdy play as substantially similar to the kinds of play of her childhood. For example, she reported that Ninja Turtles, which seemed to be universally reviled among pre-school teachers (and which Gloria also found uninteresting), was, on second take, nothing more than cowboys and Indians:

> Everything that's on television today is spun off of something that we've seen back in the '50s or the '60s. And it's just made it more vivid, more violent I think. But the basic message is the same, to me it is. You know, you see Roy Rogers hiding behind a tree or a mountain. Or you see the Ninja Turtles hiding behind a street car. They've got a tunnel instead of a mountain. They still hide. They're still waiting to surprise somebody when they come out. Or they're still defending their territory. So John Wayne and the cowboys were defending his ranch, well, they're defending their sewer or raft or wherever . . . I don't see the difference.

Gloria's house has lots of open space, broken-down furniture or no furniture at all—one reason for her calm attitude towards physical play. She also has a smaller number of children and does not feel compelled by a decorum of having a classroom space. As long as they are not 'throwing furniture across the room' or landing punches, she does not interfere. Gloria reported a much higher level of engagement in rowdy play by girls in her classroom:

> Oh yes. We've got kickbox and *Karate Kid. X-man* [*sic*]. What's that show I didn't like? *Power Rangers*. They play a lot of that. That's their karate. They really try to kick, action, do the flips like the Power Rangers. On the *GI Joe* they get down and try to hide in little places and spring out on you. Or try to do that *Karate Kid* move on one foot and all that stuff. Now I do have some girls that will get in there with them. They'll stick with them just like the boys do, you know. Try the karate stuff.

Gloria also described boys—pre-school and school age—playing with Barbies and baby dolls at her place after school (noting the boys would not do this at kindergarten or at their own homes: 'Gotta be a macho male when you go home. Dad's not gonna let you play'). While all of the other teachers we interviewed felt that popular media increased gender stereotypes, and segregated boys and girls in play, the play that Gloria described was the least gender stereotyped or segregated. 'A lot of the boys still play with dolls and get all the baby dolls dressed up . . . it's no big deal. And they can put that down, go watch *Power Rangers*, come back and pick up the baby doll.'

Gloria herself reported in her media diary a wide range of choices: videos, books on tape, books, radio, music. Her selections were eclectic in range and she seemed to have relatively little interest in 'women's genres' such as soap operas or made-for-TV movies. Her favourite movies feature Claude van Damme and Steven Seagal; her favourite shows are news magazines; she watches all sports except golf. She spoke enthusiastically about the media and the ways in which they provide her with entertainment—TV is especially important to her because she is restricted to her home / workplace seven days a week, with little or no opportunities for vacation.

Gloria's assessment of media and consumer culture were by no means uncritical. She was keenly aware of the manipulative effect of marketing strategies (such as the need kids feel to collect *all* the Power Rangers in a set). She was matter-of-fact in describing the best strategy for dealing with children's material desires:

> If you set your child down—well 2- and 3-year-olds you really can't reason with them—but 4-, 5-, and 6-year-olds, if they're giving you a hard time, you just tell them, 'Hey, this is the budget. This is what we can have. You can pick two of them out of this. This is all. That's it.'

She often lectures parents about the need to keep children away from adult films and graphic violence. She strenuously advocates keeping children away from the violent action movies she most enjoys; pulling the plug on the TV set on a regular basis; and directly teaching children the limitations on consumer spending. One of her greatest concerns is the ways parents substitute consumer goods for parental companionship. She advises parents to turn off the television set and play cards or board games with their children.

Gloria watches the media carefully for black characters: she was more perceptive and more accurate than other teachers interviewed in noting the various types of character on different shows. Gloria also noticed the changing demographics of action figures to include more black characters, even though they are rarely the main hero. She speculated that *Teenage Mutant Ninja Turtles* were popular with her children, most of whom are African-American, because they seemed to remain outside of racial categories.

When asked to envision the kind of changes she would like to see in her job, she spoke about the broader contexts of children's lives. She was the only teacher we interviewed who imagined structural changes (in the legal system, in governmental support) as a means of bettering children's lives:

> I'd probably have the government pay for everything. At a nice rate—not a low rate. And have them take some of the burdens off the parents that couldn't afford it, you know, they wouldn't have to go through so much red tape. And since I'm a foster mother I'd probably, if I ever see, well I do have some kids that are abused, I would change that system, so the parents wouldn't have them anymore. One time, you're out. If it's a little thing, I'll give them two times, but that's it. I wouldn't give them on back.

> And make sure if the kids came not dressed right, if the parents couldn't afford it, I'd have some kind of funds just out there and get what they need. If parents didn't take care of it, I'll just keep it here. If they go out they wouldn't look like second-class citizens. That's my idea. And everybody had enough to eat. A lot of my kids don't.

'No Forks in Jail, Either . . .'

Jean DeWitt has worked for four years at the daycare facility attached to the local hospital. Jean is taking college courses one at a time towards her bachelor's degree, while being employed full time at the centre and single-parenting for her three daughters, who are 4, 8, and 10 years of age. The hospital is a regional centre for medical care and one of the town's leading employers. The daycare centre is housed in a remodelled elementary school: solidly built, with a security entrance, large airy classrooms, and modern facilities. It is the largest facility in town with 138 children—a group large enough to require three staggered lunch times. The parents who use the facility include nurses, janitors, clerical staff, and physicians. Jean teaches a classroom of seven to twelve 4-year-olds, four of whom are girls.

At Jean's school, the teachers share a single monitor and video recorder on a cart and they are officially limited—by order of the school director—to two one-hour blocks of video screening per week. The centre receives cable television, making it possible to watch Nickelodeon's pre-school programming or *Sesame Street*, or cartoons, in addition to videos. This is a common practice in the 'Multipurpose' group room where breakfast is served and where children wait for their parents to pick them up late in the day. The TV being on at pick-up and drop-off times had given parents a bad impression, despite the fact that other 'choices'—toys and art activities—are always available in the multipurpose room. Some parents had trouble dragging their children away from the set at pick-up time before a show or video finished playing. Unlike Glenn, who was sympathetic to parental resistance to any TV watching at school, Jean thought TV was an important way for sleepy kids in the morning to ease into the situation of being in institutional care, and she defended the practice.

Jean voiced many common complaints about children's television and its effects on children. For example, she told this wry story about the first time she saw an episode of *Power Rangers*:

> I thought that the kids that are the Power Rangers—I thought that they set very good examples. It was like, they were upright good kids that were leaders of the school and helped out in the community and did various community-service projects. And then we were watching it here—the first time I saw it at the daycare in the morning—and we were watching it and then all of a sudden they changed into these Power Rangers and started kicking and fighting and had weapons and all the teachers just kind of looked at each other and said 'Uh' [*laughs*]. We just kind of turned it off and the kids just groaned and we knew what would happen. That they would be kicking and fighting and sure enough, they were re-enacting *Power Rangers* out on the playground, kicking and picking up sticks and pointing fingers and—

As our conversations ensued, however, it became clear that Jean held a more nuanced view of the media's role in children's lives. She held, in fact, a rather ironic sense of the media's effects and an appreciation for some of the pleasures TV viewing offered children—something closer to the cultural studies version of the interaction of TV and audiences. An avid reader of periodicals that are devoted to early childhood education (provided by the centre's director), Jean was fully engaged with the professional literature on early childhood development, while struggling to work out her own position on the best way to engage such materials in the classroom—apparently a more permissive strand of the literature on pedagogy than Montessori's writings. She gave the following humorous anecdote about attempting to employ the recommended strategy of extending the child's interest to her students:

> I've read books—the very liberal child psychology books that say 'Follow the children's lead and extend it' and one day I sat and watched some little boys play to see what their interests were. And after an hour of killing, maiming, and destroying, I thought [*laughs*], 'How do I take this a step further?' [*Laughs*] I don't think this is the direction we want to head [*laughs*].

Still, Jean has facilitated a great deal of popular culture-based play, in a manner that is reminiscent of Gloria's theme days at her family daycare centre. When a child brought in the tape of the animated film *Fievel Goes West*, some Wild West play started in the classroom. Jean made them sheriff's badges with gold glitter on them and brought in the discarded cardboard box from a washing machine to make a jail. (The box was soon appropriated by the girls for doll play.) Jean allowed the kids to see the *Power Rangers* episode 'Food Fight', and tried to incorporate it into a lesson plan about good table manners. In an attempt to downplay the weaponry, the students did a papier mâché art project making Power Ranger helmets.

Like Gloria, Jean has been an attentive observer of gendered patterns of children's play. Jean also reported a greater flexibility in gender roles—with more girls joining in action-adventure play—than did Glenn at the child development centre or Sara at the Montessori school. She observed that *Power Rangers* had brought about a qualitative shift in the kinds of superhero play going on in her classroom:

> JEAN: Well see, the girls can get in. The girls can get in. They can be more than just the victim in the *Power Rangers*, which I guess that comment lends to its credit, so to speak, as far as the females are not excluded from power. There's two female Power Rangers, so that is a switch.
>
> ELLEN: So do the girls in your class like to play it?
>
> JEAN: Oh yes. And they're not the victims anymore.
>
> ELLEN: Are they in there kickin' and—
>
> JEAN: Yeah. They've got the guns and they're right in there with them. Now when they play Ninja Turtles, they're the victim. And they'll sit in a chair and pretend that they're tied. [*Laughs*] Yeah. And they come and get rescued by the Turtles.

ELLEN: Do they play this kind of action cartoon play indoors and outdoors?

JEAN: Yes. They would play it all day long if we let them [*Ellen laughs*]. They would play it from morning—they play it in their sleep sometimes [*laughs*]. That's why it's like it's gonna come out so I just try and keep it positive. No contact [*laughs*]. And draw the line that when somebody's unhappy or somebody's getting hurt then we have to stop. It's kind of good 'cause I can hold it over their head for behaviors. Like 'I will not allow this if you can't behave this way'. So in a way it helps me modify their behavior.

Jean sees the imitative violence as a somewhat natural form of children's play rather than a long-term effect on their minds, and turns it into an opportunity for promoting classroom unity or harmony consensus. She does not worry about long-term effects of viewing media violence; she expresses a good-natured acceptance of such play as a normal phase children go through—in this, her position is quite similar to Gloria's. There is also a willingness to use overt forms of control over the children, to enact punishments or withdraw privileges. TV is useful for both Gloria and Jean as a reward that can be taken away. Neither Gloria nor Jean has any concern for the aesthetic education of the children, or for raising their level of taste to something resembling that of adults.

While Glenn and Sara exhibited a kind of blanket distaste for popular videos, Jean had personal favourites among their collections. The hospital daycare has a video library, of which musicals are Jean's favourite genre—*Kids in Motion, Wee Sing,* and *Kids Songs.* Her favourite Disney movie is *Beauty and the Beast*—because she's not a victim. Jean allows children to bring in a video from home—the rule is that they must clean up their toys in order to be able to watch it. Sometimes they make it a big party and have popcorn. As much as anything, the problem for Jean with the videos is how to fit a feature-length film into her limited screening time each week—and how to deal with disappointed children who miss parts of the show when it is broken into two segments.

Jean adopts a very liberal policy about bringing toys to the classroom, seeing the space of the school as an important bridge between home and school. Children are allowed to bring toys in at any time, but if they argue about a toy she places it in a special, brightly decorated box:

> And that box comes in handy, and it's real bright and pretty and just right there and they can see it and they know that's where all the toys go if they don't play nicely in any way really, I mean, if they don't share and take turns. So they pretty much learn as a natural consequence that 'Hey if I bring too many I'm going to lose one. If I don't share, if it's a toy I can't share, it's going to end up in the box'. They pretty much monitor their own toys and what I saw—it's happened more than once—they start bringing in two. Two motorcycles or two Ninja Turtles, two Power Rangers [*laughs*]. They'll have one for their friend, one for themselves, and they can play nicely. They do really well. I'm surprised that other teachers don't like toys from home, 'cause I think it's a very important bridge between home and school. I think you learn a lot about the kids

Girls' action figures: a bridge between home and school?

and their home life by what toys [*laughs*] they bring in. It gives you a real clue as to why a certain child behaves the way he behaves or—we gather a lot of information from the toys. And—

Jean's major concern is less for the imitative violence that Glenn worries about than for the ways that children are scared by films and TV programmes. This kind of protectionist attitude is similar to what David Buckingham found among the parents he interviewed for his book *Moving Images*: 'While parents may express the belief that *other* children copy what they see, their primary concerns in relation to their *own* children are more often to do with them becoming frightened or upset' (1996: 302). Also similar to the reports of many of Buckingham's parents, Jean views news and non-fiction as possibly more traumatic than horror or action films—in this she also resembles the parents Buckingham interviewed. Jean related a story about a child who had been watching *Cops*, a 'reality' show following police on arrests, and seemed to be behaving differently. On this rare occasion she intervened, speaking to the mother about what the child had been watching on TV. The mother blamed it on the father, and thanked the teacher for bringing it up so that she could use the teacher's authority to get the father to be more careful in his television selections.

> The teachers and I did get into an interesting discussion on how kids . . . get to watch the television show *Cops*. And we were kind of debating why parents let their kids watch this show, when it's definitely a graphic show. It has a warning before it, and, you know, these kids are obviously traumatized by what they see. I mean, it bothers them. They're upset about it. That's why they come to us and talk to us about it. And we were

> trying to . . . you know, figure out why parents let their kids watch these shows. And whether it was okay if the parent was sitting there and explaining things to them. Even then, I think it's better to shelter them. I mean, they don't need to see it at a tender age like that. Maybe, I mean it's inevitable they'll run into it, but . . . to just sit there and let them watch it.

Jean's feelings about the need to protect children are linked to her keen memory of her own childhood fears of the TV. Her father was a travelling salesman, and on Friday nights, when he returned home, the family would settle down for 'a fun TV night' in front of a programme of old horror films. Jean recalls her father laughing at her for being so scared of these 'hokey, B, horror flicks':

> And I remember being terrified and running and hiding behind his chair at these poorly made films. So I can just imagine the really good ones, or supposedly the good ones, that kids see nowadays. I'm sure they're petrified. They have to be, terrified of these.

Despite her liberal attitudes toward children's popular media in the classroom, at home Jean is quite strict about monitoring the television. She 'has read' that there is a direct correlation between grades achieved in school and amounts of television watching. She describes herself as constantly battling with her daughters over the TV set. Their family viewing is restricted to shows such as *Full House*, *Little House on the Prairie* and *Dr Quinn, Medicine Woman*—all three shows about single parents.

> Yeah, they are constantly trying to turn it on and I'm constantly turning it off . . . Saturday mornings I feel like I got to let them watch at least one. So I read while they're, I let them basically while I'm getting breakfast ready. And then it's like, 'OK, now we got to go'.

Jean's mother refused to get a television set for a long time because she thought it was better for the children to read. When they finally got a black-and-white set, Jean's favourites were *Chuckwagon Theatre* cartoons, *Cowboy Bob*, and *Wagon Train*.

> I remember I liked um . . . *Wagon Train*? [*Laughs*] I don't know who was in it but they all packed up their wagons and headed west and they had this scout leader, you know the trail leader, and the one guy that went out and scouted and, my brother and I loved that show [*laughs*]. . . . We would re-enact it out in the backyard. We had one of those little horses . . . with the springs. And um we hooked up the wagon behind it and pretend that we were headin' west [*laughs*]. Then my mom got us the little Jane and Johnny West dolls and it had the little horse and the wagon train and we'd pack it up and head out west [*laughs*].

It is interesting that Jean facilitated the western play in her classroom and reports that her daughters are interested in 'the western prairie type thing'—as represented by *Little House on the Prairie* and *Dr Quinn, Medicine Woman*. With these programmes and activities she is able to draw on her own fond childhood memories of play with her brother and western programmes, and does not have the feeling that television today is completely different from what she experienced as a child.

Jean was unexpectedly divorced when her youngest child was an infant. She had planned to be a 'full-time mom': 'I never thought that I would be a working mom, ever.' Now she was just waiting for her daughter to go to kindergarten to return to teaching full time.

> And it's the next best thing to being a full-time mom is to work in a daycare and have your child with you. It's the next best thing. That's why I say once she gets to kindergarten, then I'll be able to go back to school. 'Cause that's more important now . . . You get five years and then they're out and you can't be with them anymore . . . They need that closeness because there are so many things that are assaulting that bond between parents and children. And I think that kind of helps us keep close. I can tell how her day is going. And I don't have to try and play detective to find out what went on. I was there. I saw. And that's the number one benefit.

Five years from now, Jean would like to have finished college and gotten a job as a public-schoolteacher or a centre director—something with better pay, with insurance, and benefits.

Jean's essential philosophy has to do with involving the child in the life of the classroom, and thus preparing them for educational success. She was able to have a theory of weaker media effects because she was willing to allow that creative things can happen in these play scenarios. A contradiction seemed to emerge in the interview, however, between her relatively strict rules regulating television at home and her permissive attitude in the classroom:

ELLEN: But I'm curious about when I think about everything that you've told me, one thing that's really interesting is that you've got quite an accepting attitude toward the children's interest in TV in these, you know, shows and that kind of thing. And on the other hand, you're pretty strict with your own children—

JEAN: On the one hand, basically that's to get those kids to like school. That's the reason I put up with the *Power Rangers* in my room . . . the *Power Rangers* is definitely to make the kids want to come to school and to let them know that I'm listening to them and I hear what they like, and to get them involved. And like, 'What would you like to do at school? This is your school too.' And let them have some input and let them feel some sense of ownership that this is my school. I can, my teacher will listen to me and my teacher will respect my ideas.

ELLEN: So do you think if you were, like, communicating more negative things about their favourite TV that it would change their attitude?

JEAN: Oh yeah, they wouldn't like me. They wouldn't like school. School is not a fun place, school is like a jail. In fact, we were joking, the teachers and I, about the kids are always asking why they don't get forks at school. They get forks at home. And we were kind of saying to ourselves like, ' 'Cause this is not home, this is a jail. It's just like in jail' [*laughs*]. They don't give you no forks in jail either. I try and give them a little more freedom than that.

Power Rangers

Conclusion

What produces the stark differences among these four in their beliefs about television's effects on children? How do Gloria and Sara—both with more than two decades of experience in childcare, both highly dedicated to their jobs—arrive at such different conclusions about television's effects?

Gloria sees children as active users of television and this view is based on an acceptance of television as a normal influence on play routines. Her familiarity with genre rules on the screen and in children's play allows her to see violence as symbolic and conventional. She rates her children's cognitive skills more highly than Sara rates those of the Montessori children. Gloria expects children to be able to handle violent content on children's shows; she is confident that the children can distinguish between media fiction and reality. Her assessment is based on close observation of children's play and TV viewing and frequent casual conversations with them about the media, in the context of a warmly affectionate relationship. Gloria accepts the children's immersion in a separate peer culture and does not expect to share their predilections. She has nothing invested in improving the children's taste; she does not expect to interfere with children's play or dissuade them from their interest in popular culture. Gloria is also more comfortable with overt forms of control than Sara is: rather than attempting to 'redirect their attention', as Sara phrases it, Gloria just says 'no'. The boundaries between adulthood and childhood are clear at Gloria's Place.

Gloria's approach is in keeping with the recommendations of education scholar Anne Haas Dyson, who has studied the use of superhero stories (*X-Men*, *Power Rangers*, and the like) in literacy education at a racially integrated, third-grade classroom in Oakland, California. Dyson has articulated the position in favour of teachers' openness to popular materials in the classroom:

> Curriculum must be undergirded by a belief that meaning is found, not in artifacts themselves, but in the social events through which those artifacts are produced and used. Children have agency in the construction of their own imaginations—not unlimited, unstructured agency, but, nonetheless, agency: They appropriate cultural material to participate in and explore their worlds, especially through narrative play and story. Their attraction to particular media programs and films suggests that they find in that material compelling and powerful images. If official curricula make no space for his agency, then schools risk reinforcing societal divisions in children's orientation to each other, to cultural art forms, and to school itself. (Dyson 1997: 181)

Television is cheap, easy, plentiful, and children love to watch it. It is also pathologized—often unreasonably—by those with the most invested in status distinctions and the most at stake in professionalizing childcare. As Julia Wrigley warns, in her useful discussion of professional expertise in childcare:

> In the absence of a social movement demanding child care as a universal right, a segmented child care market will continue to provide one set of stigmatized services for the poor and other services geared to preparing middle-class children for entry into the competitive world of schooling. With such strong segregation of the children being served, caregivers can develop narrow ideologies that exacerbate the educational anxieties of one part of the population and emphasize the parental inadequacies of another. (1990: 305)

The Montessori school employs a notion of official curricula diametrically opposed to Dyson's and typical of the kind of narrow ideologies Wrigley criticized. Sara is at once completely unfamiliar with TV and absolutely certain that TV has a detrimental effect on children. Yet her classroom rules and avoidance of TV during her leisure time make her a poor observer of the phenomenon of television's effects. Sara's classroom is a place where children are supposed to get on with the work of adulthood, and this does not include the idle pleasures of television viewing. In some ways the boundaries between childhood and adulthood are more blurred at this school. 'Work' is valued over play; books and computers are valued over television and toys. Children are taught to work quietly, alone at intellectual tasks. Bothering other children with talking or gregariousness is not permitted. When 16 mm films are shown, the students are expected to hold still and remain quiet, thus a passive viewing posture is enforced. Sara's classroom is a microcosm of the environments the children can expect to experience throughout their school and adult lives as middle-class professionals. Sara openly defines her role in terms of educating tastes, monitoring and pre-selecting cultural goods. One of the most important features of her work, then, is to sharpen the distinction between children's mass culture and educational materials. Banishing television as a subject of conversation or as an activity is a linchpin of Sara's strategy for enforcing cultural distinctions.

Do rules for exclusion of media in the classroom serve to guard against the incursion of the market? It would be a mistake, I think, to view Gloria's and Jean's centres as market-saturated and Sara's and Glenn's classrooms as freer of commercial influences. In some ways, Gloria's child-rearing methods (the straightforward denial approach), her encouragement of group play rather than solitary activities, and her advocacy of traditional game-playing, such as cards and checkers, as the best way for parents and children to spend time together, are examples of child-rearing practices relatively untouched by the market. John R. Hall has noted that in some ways the working class may have more autonomy from the market than the middle class:

> The poor will find themselves exposed to the cheapest of petty commercial culture . . . And they will be the targets of culturally distinctive state- and religiously organized welfare and charity programs. These realities would suggest that the poor partake of what Bourdieu calls a 'dominated' cultural aesthetic. Yet the poor do not engage in commercial consumption in the same way that more monied popular and elite classes do . . . Paradoxically, the relatively greater distance of the poor from commercial culture will leave room for the greater importance of 'quasi-folk' cultures made in the ongoing practices of the people who live socially marginal lives. (J. R. Hall 1992: 265)

Despite Sara's and Glenn's objections to the commercialization of childhood, their classrooms are, in some ways, more invested in consumer goods and dependent on high-ticket items such as computers, CD-ROMs, educational materials, and books. Ironically, middle-class adults spend enormous amounts of money in the attempt to shield children from mass market goods and television, and to avoid advertising which directly targets them (Seiter 1993). The consumer goods available for children's use at the Montessori school are more controlled by adults in immediate proximity. While the argument is often made that children should be kept away from television because they cannot understand it, in fact it is often the ease with which they become media experts that may frighten adults.

The Montessori school represents the type of institution afforded status by those class fractions predominant in the academy and most similar to the petit bourgeois intellectuals at the centre of Bourdieu's study in *Distinction* (1984). But it would be a mistake to see Sara's system of cultural representations as the only legitimated one in this community, recognized by one and all as the superior system. In Bourdieu's model, distinctions are used to legitimate the privileges of those with more education and more money, who envision themselves as superior to those whose tastes differ from their own. But John Hall has criticized Bourdieu's model as too holistic, and too rigidly tied to class rather than cultural (and subcultural) distinctions: 'Cultural distinctions do not represent some generalized currency of "legal tender" among all individuals and status groups . . . Cultural capital, after all, is good only (if at all) in social worlds where a person lives and acts, and the value that it has depends on sometimes ephemeral distinctions of currency in those particular social worlds' (1992: 275). This incommensurability of cultural distinctions may be found in the childcare market. While Sara's and Glenn's classrooms offer the kind of cultural capital valued by the educated middle class, it would not be appreciated by all parents. Some parents value experiential knowledge of child-rearing and the emotional availability of the care-giver more than the kind of formal curriculum Montessori offers: these are the kinds of asset Gloria offers, along with the simple fact that her hours of operation accommodate the many workers who are not on a nine-to-five schedule. Beliefs about media effects on children are inextricably bound to adult use of the media, class position, and ideologies of childhood. Sara loved television as a child, but views contemporary television as completely different from the *Lassie* that she once enjoyed. She completely avoids television in preference for higher status media forms—cinema, radio, art films. Similarly, Glenn carefully moderates his own television viewing in favour of more worthwhile activities. Or, at least, this is what was self-reported.

Gloria's and Jean's interviews suggest that a less deleterious view of media effects may emerge from situations where adult care-givers know more about the media, invest less in status distinctions, and create an environment where children feel free to talk about media without inviting adult disapproval. Their freedom from persistent anxiety about media effects allows them to be more indulgent with the children in their care. These women have arrived at very similar strategies despite very different patterns of media usage in their own lives: Gloria is an enthusiastic TV viewer and an avid fan of action-adventure films, while Jean has scarcely any time for TV viewing at all. Jean is less indulgent with her own daughters than with her students, however, suggesting that there is a gap between her professional and her parental views on media effects. Paradoxically, Sara's and Glenn's students, who live in the most affluent and protected homes, are viewed as gravely at risk of deleterious media effects, and in need of constant protection from the influences of media and consumer culture. For adults as well as children, holding a lay theory of deleterious media effects, and the alarmism that accompanies it, may be linked to the closeting of popular culture tastes, and to the suppression of talk about television. Ironically, middle-class adults—exceedingly anxious about the collapse of the public schools, and the diminished opportunities for upward mobility through education—may be the harshest and least empathetic about children's interest in popular culture, while those with less money and fewer chances of advancement may have the luxury of being empathetic to children's interest and enthusiasm for popular toys and television shows.

In the next chapter, I examine the critique of the mass media offered by Christian fundamentalists, to discern how this form of 'lay theory' parallels as well as deviates from the secular scares about mass media effects (Gilbert 1978) and the pronouncement of childhood experts such as psychologists and educators. With renewed concern about the increase of violence on television, the lack of content regulation on cable, and the weakening of Standards and Practices Departments, strands of the political right and left appear to be coming together in their advocacy of renewed censorship. As historian James Gilbert usefully reminds us, the concern over the mass media's corrupting influence on children, its weakening of parental authority, and its ability to inculcate values, is an 'episodic notion': one that could be introduced at any time and thus must be studied for its social function, and how it inevitably reflects and shapes social relationships. The problem with the old worry about mass culture's effects on children, is that 'the control of youth is, even in the best of times, problematic, and . . . the cultural history of the United States is filled with dramatic changes in the content and form of popular culture' (Gilbert 1978: 4). Thus, the timing of public concern over media's effects on youth, and the ways it is introduced into secular, political, and religious discourses, warrants careful scrutiny.

5

TV Among Fundamentalist Christians: From the Secular to the Satanic

Ellen Seiter and **Karen Riggs**

For Christian parents the struggle to protect their children from the culture goes far beyond junk food and celebrities pushing sneakers. It has become a daunting task to shield the younger generation from 'safe-sex' instruction in school, from profane and sacrilegious language in the neighborhood, from immorality and violence on television, from homosexual and lesbian propaganda and from wickedness and evil of every stripe. How unpleasant it is to continually fight for decency at home.

Reverend James Dobson, Focus on the Family Newsletter, *1994*

THE MAINSTREAM, middle-class critique of commercial children's television is well known: it promotes aggressive behaviour, unhealthy eating habits, and disputes between parents and children over spending on tasteless toys and junk food. Fundamentalists offer a different critique of children's television and consumer culture, based on perceptions of Satanism, magic and supernaturalism, New Age themes, Darwinism, and taboos on sexuality—especially queer sexualities—and nudity.

In this chapter, we compare the perspectives of teachers employed by fundamentalist, working-class churches and fundamentalists running home daycare centres. How do conservative Christian women view their work with small children and the significance of their role as media gatekeepers? Are fundamentalists' views distinctively different from those of others doing such work in secular settings? The five case studies presented here offer fascinating examples of oppositional readings of the mass media, but from a conservative Christian fundamentalist position, rather than the proto-feminist and socialist readings often implicitly sought in the cultural studies tradition.

The community we studied is one where much of the available daycare is organized by churches, and thus one component of the special benefits of church membership for mothers is access to childcare. The town has a population of about 50,000 and is close to the central states' Bible Belt. Revivals and camp meetings are advertised in the local newspaper on a regular basis;

Bible-study groups are an important opportunity for socializing for many women. For these reasons—and out of an awareness of the growing national prominence of the Christian Right's critique of TV and the emphasis on harmful media effects on children propagated by conservative evangelical groups such as Focus on the Family—we deliberately sought to include in our sample as many teachers associated with fundamentalist churches as we could. We wondered how different women associated with fundamentalist churches were from the other women we met during the study. How did they conceptualize the media's effect on children? Was their leisure time media usage very different from that of other women employed by secular pre-schools and daycare centres?

Before moving on to a discussion of the various forms of Christian fundamentalism and to a review of the pertinent research on media audiences, we wish to situate our discussion in terms of a larger project of work on the Christian Right. Along with other feminist researchers such as Rebecca Klatch (1987), Judith Stacey (1991), Linda Kintz (1997), and Kintz and Julia Lesage (1998), this work represents an attempt to rethink one category of feminism's 'others': women who are working class or lower-middle class *and* deeply involved in fundamentalist churches. In doing our research, we were immediately confronted by the prominence of class differences in our sample, and impressed by the need to view religious affiliation in this context. From the outset, our field research challenged us to imagine the context in which working for $5 per hour in a church-affiliated nursery would seem to be a good job, a job to be grateful to God for having, as many of our subjects felt it was. From our vantage point as college professors, this employment scenario seemed distinctly unappealing.

Julia Lesage has pointed out the ways in which working within church groups and Christian Right coalitions can hold out a promise of upward mobility for women with little access to job training or education. For these care-givers, working for a church nursery offered an ability to 'be one's own boss' in the classroom. While not lucrative, being a pre-school teacher represented a gain in status for some women, moving from unpaid homemaking or from custodial work. In terms of the church, the job offers insider status within the church community, and even a divine mandate for one's work, in the sense that it involves doing 'God's work' with little children. Finally, this is a job where women can bring their own small children with them to work, monitor what is happening to them at daycare, and procure childcare for free or for reduced fees.

In a number of interviews, women related their work and their church participation to a desired level of respectability (Skeggs 1996) that may otherwise have been difficult for them to achieve as working-class people. Of the five, only two held college degrees; only one had a family income of more than $25,000 a year. Sealed off from high levels of education or high-paying jobs, these people found in church involvement the promise of respectability and, at times, an alternative to unemployment. Members of a congregation often perceive participatory roles as prestigious; to be in charge of the altar guild can make one part of a sort of elite.

Establishing respectability may seem like an urgent priority for working-class women at a historical moment when the economic and social safety net is being ripped away from women and children. Respectability in the church may seem to promise protections from attack, from poverty, from violence, from unwanted pregnancy. As Helen Hardacre usefully reminds us, there is an 'element of economic realism' in the way in which women turn to fundamentalism, which is tied to a conscious desire to secure the husband's loyalty and economic support during times of a 'no-win game on the labor market' (1993: 142).

One means of establishing respectability is to demonstrate seriousness about the moral and religious education of one's children. Issues of child-rearing and media censorship are today among the most powerful draws the Christian Right has to offer. The runaway hit of Christian broadcasting, in the wake of the Bakker and Swaggert scandals and the fading of Falwell and the Moral Majority (Pettey 1990), is the Revd James Dobson, a Ph.D. in psychology, a licensed therapist, and a former associate clinical professor of pediatrics at the University of Southern California. Dobson parlayed a radio station and a string of books on child-rearing (with titles such as *The New Dare to Discipline* and *Parenting Isn't for Cowards*) into a 'family ministry' with an operating budget of $66 million a year and carriage on 1,000 stations (Shumate 1996). His formula for success is based less on prayer and Bible study than on the conventions of popular writing by child-rearing experts and critics of mass media's effects on youth—which gained particular force in the 1950s (Gilbert 1978). Dobson's themes, like those of many other psychologists, psychiatrists, and experts on Christian radio, focus on 'how to strengthen marriages, improve relationships, "build hedges" against the temptations that lead to sexual sin and keep children away from the ungodly "world" ' (Shumate 1996: 9).

Dobson's explicit mission is to offer spiritual advice and counsel to parents struggling to rear Christian children in a world fraught with evil. His newsletters and broadcasts repeatedly emphasize popular culture's corrupting influence on children, spreading homosexual themes and 'gender feminism'. Heather Hendershot (1995) has noted that the conservative Christians' crusade against the damaging effects of media is one of their most powerful weapons, and Dobson encourages his listeners to engage in close readings of media texts to discern the meanings nefariously hidden in song lyrics and television images—and toys:

> Even today's toys seem to have been designed by politically correct manufacturers. Remember, for example, the Barbie and Ken dolls that have been popular with several generations of children? Well, take a look at how Ken has evolved. In the latest version, he wears symbolic jewelry, including a necklace of rings and articles of clothing that hint at his 'sexual preference'. He might not be interested in Barbie at all, if you know what I mean. (1994: 2)

Using a popular strategy of the Christian Right, Dobson targets specific shows, ranging from *Beavis and Butthead* to *Sesame Street*:

> Many US television series have included episodes promoting the Gender Feminist agenda, homosexuality and choosing one's own sexual identity. These include the popular *Roseanne*, *Northern Exposure*, *Star Trek: The*

> *Next Generation* and *Star Trek: Deep Space Nine*. The message that sexual identity can be deconstructed and the vocations of manhood and womanhood are nothing but socially constructed gender roles can be found in children's books and on the popular children's television series *Sesame Street*. (Dobson 1995: 1)

Dobson does not restrict himself to advice on running the home and family. In the tradition of evangelicals who strive to make an impact in the public sphere, his activist right-wing political agenda is prominent in all his materials, and he gets big results. Dobson mobilized his listeners to make nearly a million phone calls to Congress in February 1994, to squash a legislative amendment that he had portrayed as a move to outlaw Christian home schooling.

Similarly, Donald Wildmon's American Family Association (AFA)—originally formed in 1976 as the National Federation of Decency (Abelman 1990: 275)—monitors television using the categories: 'anti-Christian; promotes homosexual agenda; profanity; objectionable sexual content, substance abuse or violence' (*AFA Journal* 1996: 6). The AFA organizes its membership, which Wildmon estimates at 500,000 (Rubin 1996), in letter-writing campaigns and boycotts, to protest 'the filth on TV and the effect it is having on our children and families'. Wildmon has portrayed an image of the family under urgent, grave threat in books with titles such as *Home Invaders*. The Walt Disney Company incurred his greatest wrath when it 'extended company insurance benefits to the live-in partners of homosexual employees' and 'allowed homosexual celebrations in its theme parks' (*AFA Journal* 1996, 'Dirty Dozen'). Disney's children's movies are the subject of outrage and cynicism: 'A priority of Disney's current philosophy of movie-making seems to be "pushing the envelope". Roughly translated, that means including as much sexual content as possible and defending it with high sounding arguments about artistic integrity, but still having the public buy it in huge numbers' (*AFA Journal* 1996: 9). In addition to boycotting Disney—an understandable target given the size of the Disney organization, its family focus, and its ability to gather publicity for Wildmon's campaign—Wildmon's hit list includes such giants of children's product marketing as Johnson and Johnson, Procter and Gamble, McDonald's, Mars, and K-Mart.

Focus on the Family and the American Family Association are just two of the many right-wing groups targeting families with children, which also include the Traditional Values Coalition, the National Association of Christian Educators, Phyllis Schafly's Eagle Forum, the Family Research Council (which is the Washington think-tank affiliated with Focus on the Family), and Concerned Women of America.

The themes of Dobson's and Wildmon's rhetoric—as well as those of many other fundamentalist groups—are quite close to the complaints voiced by middle-class educators (such as those discussed in the last chapter) about children's media. The fundamentalists are much more explicit than the secular educators in their condemnation of sexual themes or a perceived lack of modesty; and they are much less explicit than the secular educators about issues of violence. Both groups agree that there is too much marketing aimed at children and too much inappropriate material on television. The difference is that an activist solution is proffered by these religious groups—letter-writing, product boycotts—rather than the individual solution of parental censorship of TV programmes and videos and the refusal to purchase certain consumer goods for children.

Twentieth-Century Christian Fundamentalism

The fundamentalist childcare-givers struggled to mediate their position between the top-down message of their church leaders ('Avoid the devil's temptations') and their own rootedness in everyday American, working-class culture. While they tended to feel guilty about their own media consumption, whether it meant watching *Roseanne* or letting their kids watch *Power Rangers*, they were, first and foremost, people without much money and without access to many activities outside the home other than church. As parents of small children, they were embedded in secular profane life through the children's consumer desires and identification with peer culture. Watching television and videos was, with few exceptions among them, the major option for adult, child, and family entertainment; although many of them reported reading the Bible, alone and with family members, several times a week, most of them tended to spend more time watching television.

All the care-givers we interviewed share membership in *fundamentalism*—one component of the broader context of conservative Christianity. About half were fundamentalist non-Pentecostals. The others belonged to churches that embraced both movements. The dominant branch of Christianity in the United States in the twentieth century has been what has been termed 'mainline Protestantism', and it has included such sects as the United Methodist Church and the Presbyterian Church (USA) (Carroll and Roof 1993). This mainline branch has been marked by a largely moderate position with respect to such issues as the literal interpretation of the Bible and tolerance of social difference. In recent decades, membership in mainline denominations has fallen off, but membership in socially and biblically conservative churches—along with the political influence of these organizations—has grown (Kelley 1972; Hunter 1987). One of the appeals of conservative churches has been their link to *evangelicalism*, the basic tenets of which are a driving interest in conversion, biblicalism, activism, and crucicentrism (Gordon 1991)—a tendency which pushes towards interventions in the public sphere, or towards issues of national prominence, such as a call for increased media censorship. Evangelicalism is neither inherently conservative nor inherently liberal, but many evangelicals are linked with religious and political conservatism, partly because of the movement's Bible-centredness. Two of the most influential conservative evangelicals have been the Revd Dr Billy Graham and the Revd James C. Dobson.

Out of the evangelical movement has grown contemporary Christian fundamentalism. Some scholars place fundamentalism under the umbrella of evangelicalism, but some cleave it from the larger movement (Marsden 1982; Hunter 1987; and Ellingsen 1988). One of the defining distinctions is that evangelicalism does not pointedly distance itself from non-members, while fundamentalism works to achieve separation on many fronts from non-fundamentalists. The more militant fundamentalist groups make proscriptions against association by their members with non-fundamentalists (Ellingsen 1988).

Fundamentalism is a label rejected by many conservative Christians because of its perceived lack of middle-class respectability and its links to controversial media figures such as Jerry Falwell and Jim Bakker. Nevertheless, fundamentalism forms a dominant strain of the Christian Right. The conservative Lutheran Church-Missouri Synod, while it shuns the label 'fundamentalism', espouses fundamentalist beliefs, such as an assertion that the Bible is the absolute word of God. The Christian Coalition, taking up the cause of fundamentalism explicitly in the national arena, inflects right-wing politics with fundamentalist-based views on such issues as abortion and homosexual rights.

Fundamentalism is a movement, not a denomination. Some denominations are intrinsically fundamentalist, such as the Southern Baptist Convention. Some fundamentalist churches are non-denominational, as were most of the ones we encountered in our study. In the southern Midwestern city where we located our sample of childcare workers, Christian fundamentalism largely was served by non-denominational churches; in the South, many more fundamentalist churches can be found among established denominations. Evangelicals (and fundamentalists) are often working class, but some are middle class (Hunter 1987).

Apart from fundamentalism is charismaticism, whose defining characteristic is a belief that God's work in the world is real and should be testified to. The most prominent strain of charismatics are Pentecostals, noted for their belief that God works in them through the Holy Spirit, and this often means speaking in tongues. Pentecostals apparently belong most often to the working class. This movement is not directly tied to the politics of the Christian Right (Quebedeaux 1976). In fact, many African-American churches are Pentecostal but not fundamentalist. Fundamentalists and Pentecostals are not mutually exclusive.

One of the distinguishing characteristics of conservative Protestants has been their child-rearing philosophy, which has placed greater emphasis on obeying authority than on intellectual autonomy (Ellison and Sherkat 1993). Drawing on data from the National Survey of Families and Households (1987–8), social scientists Christopher Ellison, John Bartkowski, and Michelle Segal found that conservative Protestants spank their children as punishment more often than other parents do (Ellison *et al.* 1996). As a major reason for this, Ellison *et al.* offer the explanation that conservative Protestants stress the doctrine of biblical inerrancy, a doctrine that posits children as naturally sinful creatures who must be set right by their parents. While the dominant child-rearing literature in the late twentieth century may stress a doctrine of building self-esteem, the conservative Protestant view—exemplified by such authorities as Dobson—is that children should be reared with the goals of salvation and conformity to the Christian life (Bartkowski and Ellison 1995). The findings of these sociologists resonate with our impressions of the fundamentalist childcare-workers' views of their own role. We understood these people to see themselves as parental-type authorities, guardians of the children's souls. Much of what they told us about their influence on children's television viewing had to do with keeping the children on the straight and narrow morally, keeping them away from temptation. It was assumed that the children themselves were inherently corruptible and in need of constant redirection.

Our interest in this chapter is in connecting particular decodings with expressly held ideological positions—in this case, religious faith. Relatively little research exists on the role of religious belief in audience interpretations. Stewart Hoover's (1988) qualitative work with viewers of the *700 Club* analyses individuals' attraction to the show based on 'faith histories' —involvements in a variety of religious organizations from childhood to adulthood (1988: 17). Hoover's *700 Club* viewers tended to have evangelical backgrounds and were often conscious of class distinctions among forms of religious broadcasting, thus the programme was appealing as a more up-market and more intellectual version of fundamentalism. Press and Cole (forthcoming) gathered distinctively different—even diametrically opposed —decodings by pro-choice and pro-life women of the made-for-television movie *Roe vs. Wade* and a *Dallas* episode dealing with abortion. Our research maintains a focus on viewing under specific material circumstances—something often lost in focus groups—and an emphasis on the relationship between media consumption and various forms of labour, such as salaried childcare and unpaid domestic work, while linking beliefs about media effects with religious beliefs about sin, the dangers of paganism, and the importance of divine—as well as parental—authority.

Introducing the Care-givers

Among the people we interviewed, most seemed to feel that their church-related pre-school or childcare work allowed them to perform a significant public role. This linkage came through most saliently in the experience of Missy Jones, the most plainly 'working class' of our interview subjects. She and her husband had both worked as custodians on the graveyard shift at the local university. She was forced to quit her job towards the end of her second pregnancy:

> When I quit my job [at the university], I knew I was going to come to work here. My husband even said, 'You know, I feel like God really does want you there.' And so we were all prepared and then it didn't happen for a few months. I only make $4 an hour, but I really feel like God wants me here. And a lot of people, if they're not religious, wouldn't understand that. But I really feel like he did, because it was just so strange. We really needed a little bit of extra money to buy some, just like groceries, that's what it always seemed like we come up short on. And this job down here just opened up, and it's just like it was totally amazing. The girl just quit one day, and it was unexpected, and they needed someone really fast, and they couldn't find anyone. And so the director prayed that God would send the right person, and she said that that afternoon she called me about just volunteering about something in the church and she said, 'By the way, you don't need a job, do you?' And I said, 'Oh, yes, I do! I've just been looking for just a little something.' Plus I can bring my kids, because if I got a full-time job somewhere and tried to pay childcare for two kids, I'd probably be better off to stay home.

Missy said she realized the director of the pre-school was watching her in the pew in church. Later, when offering the job, the director told Missy that she had impressed her. About that time, Missy's husband became the church youth director, and she said he hopes to go on to the ministry himself. In non-denominational churches with working-class congregations like this one, pastors are often not highly educated; in such churches, the ministry might seem like a more obtainable goal to a janitor and his wife than it would in a church where the pastors have graduated from seminary.

This sense of performing a public role came through to varying degrees among the care-givers in our sample. They included:

1. Janice Root, aged 35, who lived in a trailer park on the outskirts of town. Her husband, a heavy-equipment operator, was, at the time of our interviews, laid off for much of the winter. Three of the couple's four children lived at home. They ranged in age from 7 to 11. A 19-year-old stepson had moved out on his own. Before pre-school work, Janice had worked as a homemaker, home-schooling two of her children, and a legal secretary. She had some college education, but had not graduated. She earned $90 a week for her part-time position and the family had an income of less than $25,000 a year. She and her husband had become progressively active in Clarksboro Christian Church after she became a teacher there a few years prior to our interview. The church was non-denominational, fundamentalist, but not charismatic. Since becoming a pre-school teacher, Janice had decided, with her husband, to plan on doing missionary work in Africa.

2. Susan Gregg, aged 33, was a member of the Pentecostal Community Church, where she was Beth Mueller's (see below) son's Sunday-school teacher. She, her locksmith husband, and three sons were active members, although her grown stepson was not. Susan worked with Janice Root at Clarksboro Christian Church. She had a bachelor's degree and, like Janice, earned $90 a week. Her family income was between $26,000 and $50,000 a year. She lived in a tract-style, working-class neighbourhood of homes of about ten to fifteen years old in rural Clarksboro, near the university city where we did our research. Susan had lived in the area for twelve years, having lived in larger cities within the state. Her three-bedroom home was meticulously ordered and decorated with needlework and china pieces. The family television set was in the small living-room, which opened out onto the kitchen.

3. Missy Jones, aged 25, and her husband both worked as custodians on the third shift at the local university before Missy worked at Hope Crossing, a pre-school in the basement of Midway Church, a fundamentalist, non-denominational, and non-charismatic church. Missy had attended some college, but did not graduate. Her $160 a week income was part of the family's less than $25,000 annual income. She had lived in the town all her life, and she and her husband had two young sons.

4. Peter Taylor, aged 37, and his wife had their sixth child during the period of our research. The three eldest were being home-schooled at their house, where the couple ran a neighbourhood daycare centre. The house, in a working-class neighbourhood just outside the city limits, was about fifteen years old and had three bedrooms. The couple drove a new mini-van to their frequent visits to the non-denominational, fundamentalist, charismatic

church known as Garden of Gethsemane Fellowship. The church had no permanent structure, but met in the gymnasium of a private school in the city. Peter's wife had a serious heart condition and had to have frequent, regular rest periods. Partly as a result of this, the responsibility fell to him to perform a relatively large number of the household tasks, such as cooking and bathing the children in the evenings.

5. Beth Mueller, aged 36, lived in married-student housing with her husband, a doctoral student in music, and their two sons, a kindergartener and a 3-year old. The family was active in the Pentecostal Community Church, a non-denominational, fundamentalist, and charismatic church. Beth attended Sunday school and church services with her family and a weekly morning prayer group for women. She made about $125 a week doing childcare in her small apartment. Beth grew up in a small Midwestern city and lived for several years in Honolulu, where her husband, a percussionist, was in the Honolulu Symphony. She had a high-school education and had worked full time as a homemaker since marrying. The family had a small television set in the apartment's living-room.

Conflicts over Media Use

In many ways the fundamentalist women we studied were indistinguishable from the care-givers we interviewed in secular settings. For long parts of each interview their descriptions of children, their classrooms, or care environments, and their concerns about media effects were in no way distinctive from those expressed by Sara Kitses, Glenn McClintock, Gloria Williams, or Jean DeWitt. Christian pre-schools appeared to adhere to the community norms for the appropriate curriculum, such as devoting each week to a letter of the alphabet, developing an appreciation of live performances, and providing arts and crafts activities. These schools added a component of Bible stories, used videos distributed by fundamentalist groups, and, in the case of one programme, gave the children a biblical verse to memorize each week. But these activities tended to take place against a background of Disney videos, Barneys, Barbies, and Power Rangers brought in for show and tell that was remarkably similar across childcare institutions.

The interviews did bear to varying degrees the traces of a distinctively fundamentalist interpretation of popular media texts, where care-givers singled out sexual imagery, magic, Satanism, and supernaturalism as special categories for concern. Some of these concerns are not unique to conservative religious thinking. For example, concerns about sexual imagery in children's media texts is not only a conservative Protestant issue, but a secular feminist issue. But we contend that the motivations for these conservative Protestants are specially inflected by their concern that the authority of the Bible be submitted to. Female sexuality, for instance, is framed not in terms of the political field of gender relations, but of a preference for the wife–mother role over that of the unmarried sex partner. Concerns about the Disney fantasy are not framed by these conservative Protestants in the terms of the typical liberal critique that females and minorities are caricatured, but in terms of the heresy of suggesting supernaturalism outside the realms of the Deity and the devil. So, we maintain, while many of the criticisms these care-givers raise are common to the criticisms of the mainstream middle class and of the secular care-givers we interviewed, the basis for these critiques is often different.

For example, these providers show a special concern with the role of Barbie dolls in children's play. This theme is also found in Focus on the Family newsletters, as well as the books of Beverly LaHayes, who has argued that Barbies 'encourage little girls to think of themselves as sex partners instead of mothers' (quoted in Lienesch 1993: 84). Hope Crossing's Missy Jones said she discouraged children from bringing in Barbies because the children tended to undress the dolls. When asked what kind of dolls she preferred, her answer bore the traces of shame about the female body: 'Definitely the ones with their clothes on. This seems kind of innocent in a way, to me, just a plain old doll.' She recalled her own simple childhood doll, which was based on Mrs Beasley, a favourite doll belonging to a character on the TV show *Family Affair*:

MISSY: Even if you took her clothes off, you wouldn't see anything. It was just always kind of innocent. She was always my little friend or something when I didn't have one. That's why maybe I feel like dolls can be innocent little toys, and they can also be their friend as well. But I'm sure a lot of the people would probably object to, matter of fact, I know they object to us having Barbie dolls in here. One day I was thinking about buying some Barbie dolls, just for the girls in the class, 'cause they really like them. But then we [teachers] were talking about that, and we were afraid, because their clothes come off, you know, and that might make them ask questions. So the more I thought about it, you know, I don't want to get myself in a situation like that.

KAREN RIGGS: Who do you think objects, the parents or the school?

MISSY: The parents, I would say. It just seems like anymore, innocent things can be turned into something horrible. I would never put myself in the situation to get in something like that. I care about these kids too much. But I was just thinking more so on the innocent terms.

Missy's answer is evidence of an enormous anxiety about her new responsibilities as a teacher, and the possibility that sexuality might somehow emerge in the classroom. She is deeply torn between her own memories and faith in childhood innocence and a potential 'situation', 'something horrible' that she can hardly bear to put into words, but seemed to have to do with sex play among children, or possible allegations of abuse from childcare-givers.

Peter Smith, on the other hand, who came from a rural, fundamentalist and very poor background, seemed aware of but refuted the fundamentalist critique of Barbie as sexualized, and was able to incorporate her harmoniously into his family daycare. When we asked him to describe his children's Barbie play, he told us that the scenarios reflected the values and activities of their home life: 'Barbie goes to church. Barbie's married and has a family.' His placid acceptance of Barbie as an innocuous toy exemplifies the possibility of rejecting fundamentalist positions in favour of the familiar, accessible consumer culture of working-class life. Peter seemed both to have more confidence than the women in his own ability to decide the moral value of toys and videos, and to be free of the kind of anxiety over sexual themes that plagued the women childcare-givers.

Another special concern of the fundamentalist providers is supernatural themes in children's television shows and play—a theme that has interesting parallels with the worry of Montessori educators about children's failure to distinguish fantasy from reality. At Clarksboro Christian Pre-school, for example, the Halloween holiday, because of its concern with supernatural creatures, offended some fundamentalist parents, and the school switched to a cowboy and cowgirl theme for the occasion. Janice Root said the people who 'raised a stink' about Halloween were her friends, and she treaded the territory nervously when these friends were around. With some irony, Janice noted that changing from Halloween to a cowboy / cowgirl party merely traded one problem for another: what to do when the dressed-up pre-schoolers effectively break the school's anti-weapon rule by packing guns for the party.

Janice's co-worker Susan (who belonged to a Pentecostal church) seemed more concerned about the role of the supernatural in child's play. Obviously, tensions existed between these two women (both fundamentalists, but only one Pentecostal) over the strictness of their censorship of children's popular culture. Susan confessed that she had allowed her own children to watch *Power Rangers* without realizing what they were in for. She did not let her children watch the programme again after she saw a 'lady casting a spell and evil stuff, and it just really didn't seem like a real positive thing for my kids to be watching'. Susan had been especially surprised because the programme had been suggested by a visiting friend from the Christian pre-school, and she noted disapprovingly that the child's parents were not monitoring his television viewing.

All the care-givers used Disney tapes on a regular basis where they worked. Among the secular teachers, there was a surprisingly high level of critique of Disney films for gender stereotyping and for violence. The fundamentalists' discussion of Disney was distinctive in its focus on supernaturalism. Susan and another care-giver, Peter Taylor, both brought up concerns about the Disney film *Fantasia*'s portrayal of demons and evil (another theme voiced by Dobson in a Focus on the Family newsletter). Peter summarized his objections to the film:

> There's somewhere in the, at the end of the movie, they have this guy who's representing evil. He's got horns and wings and he comes out of the ground and he calls up all the dead souls and he's really huge. He's really dark. All these skeletons are flying through the air and on horseback and things like that, and they're chasing people around . . . And he's like casting lightning bolts . . . It's scary for kids because, maybe more so for even my own kids, because for them the reality of good and evil is a reality, you know. And to see that, because I think what that does is it gives too much power and authority to an evil presence. You know what I'm saying? It's like, OK, sure, the devil may have power, but his powers are limited, as much as God will allow him to have.

In *Fantasia*, Peter added, the devil appears 'all-powerful' and he did not want his children to see that image because 'there's nothing to counter that in the movie'. Missy Jones said she often showed her students cartoons featuring Porky Pig or other characters from *Bugs Bunny*, but, when a child brought in a tape featuring the Tasmanian Devil, she had said, 'No, I don't think we need to watch this today.' Although she had never seen the tape, she recognized it as unsuitable based on the title character's link to Satan.

Some of the teachers voiced an inner conflict over their tendency to approve of some supernatural content while spurning others. Susan, for example, said she thought such portrayals of magic as Disney's *The Sword in the Stone* were harmless, because it is a 'cute movie' and the magic's fiction is obvious. On the other hand, while she found Barney to be a 'pretty much OK guy', she was bothered by the show's easy portrayal of magic. She said that children should not have the opportunity to be desensitized by the presence of sorcerers and witches on shows that treat magic as harmless. However, she said she liked Barney's encouragement to children to use their imaginations. Unlike Janice, her colleague, who decorated her home with Jack-a-lanterns at Halloween, Susan tended to hold negative feelings about the holiday because of its glorification of evil spirits. Acknowledging that she had trick-or-treated as a child herself and it had done her no apparent harm, she said she allows her children to do so as well, but does not let them dress as witches, ghosts or other scary creatures. She felt guilty about allowing them to participate in a 'heathen' holiday, but did not want to deprive them unfairly of the candy that their friends were surely collecting.

Peter Taylor said he used *Barney and Friends* tapes in his own daycare out of necessity, but with a marked unease. He said he showed the popular programme to children, including his own, but less often than he showed *Sesame Street* and *Mr Rogers*, because *Barney* seemed 'too magical':

> I think it's the way they will chant to get something to happen or, you know, they'll say these magic words and they don't so much associate it with imagination. Let's do this and this will happen and, you know, I mean it's different than Mr Rogers saying, 'Let's go to the land of make-believe.' You knew that you were in the land where people were dressed up in costumes. There's a difference there, at least as far as I'm concerned.

Mr Rogers was Peter Taylor's favourite children's television show, and there seemed to be a close identification with the image of the kindly paternal figure that made questionable content safe. We suspected that Peter might be emulating Mr Rogers's style with children, perhaps thinking of work with children as a kind of ministry. Still, Peter said he was drawn to Barney's teaching, because he liked the way the programme imparted pre-school skills such as counting and rhyming.

Missy Jones also had some conflicting feelings about Barney's supernaturalism. She said she tended to show *Barney* tapes to her 3-year-old class a few times a week, partly because of their popularity with the children, who frequently brought them from home to share. When we interviewed her, however, she had recently begun to show the tapes less often:

> This one parent is having a problem because the child was saying that Barney is better than Jesus. And they're like, you know, 'You've just got to help us. You've got to help explain that that's not so.' And so I did, and we just kind of cut *Barney* off a little bit around here. We were watching a lot of *Barney*.

Missy may be familiar with the critique of *Barney* circulated through a sermon booklet published by a South Carolina minister in 1993, which argued that the programme 'is immersed in New Age messages', 'constantly teaches transcendental thought and mystical ideas', and 'everything on the liberal left's agenda from New Age evolution to radical ecology' (Chambers, 1994). Targeting a book in the *Barney* series about a little girl's friendship with the dinosaur, Chambers wrote 'Clearly Barney has replaced her parents as the source of her peace of mind and stability . . . It certainly leaves no room for the lord Jesus Christ to be a child's best friend.' Although Missy seemed to disagree with this type of critique, and wished to defend *Barney*, she said it was difficult for her to get her message across: 'It's just hard because I feel like I just want to stand up and say, Barney's not really gonna do all this, but I don't say that. I just try to keep on teaching what we're supposed to teach and then tell them that Barney is good but Barney's not everything.'

Missy's ban on Barbies and Barneys from her classroom is typical of the boycotts urged by various Christian Right organizations, and is an example of the hostility to corporations found in some strains of fundamentalism. On the other hand, the Christian Right is vigorously promoting its own form of marketing in the form of home schooling materials, Christian book clubs, videotapes, musical recordings, toys, and games.

Leisure Viewing by Care-givers

In the course of our interviews, it became clear that women used church attendance as a means of stimulating social life—in ways not afforded by private, domestic television viewing. They were more likely to be going out during the week with their husbands than some of the other married women in the sample. (Because we conducted these interviews in winter and because incomes were low, most people were house-bound with little but TV to turn to for recreation.) They frequently discussed the joint nature of their parenting activities, referring to fathers as active partners in decision-making for children, and in some cases with childcare. One woman even indicated that she got to watch whatever she wanted on TV because her husband was laid off and taking care of most of the housework. This family was reminiscent of some of the working-class families studied by Arlie Hochschild, who found that, despite an official family version of gender relations that argued that housework and childcare were women's work, working-class men were more likely to be shouldering the burdens of domestic labour than their middle-class counterparts, although middle-class families tended to have express ideologies of gender equality (Hochschild 1989).

Family television viewing, beyond programmes purely for children's viewing, varied widely among the fundamentalists we interviewed, with Peter Taylor's and Missy Jones's families being the most conservative about programme content selections. Almost all of the providers, in fact, said they tended to be more conservative about their family viewing choices than their parents had been, because television programmes, for children especially, had grown more violent.

Peter told us that by the time he finished making dinner at the end of the workday, feeding the children, and sharing bedtime preparation responsibilities with his wife, the couple were exhausted and had little interest in television for themselves. Also, because of their church activities, they were not home two or more evenings per week. He said he and his wife frequently read the Bible in the evening.

In her media diary, Beth also reported Bible-reading as a primary form of media consumption. She often read the Bible alone in the morning or in the evening with her husband and children; frequently, the family read stories together from its children's Bible. Missy reported that she generally read the Bible by herself in the evenings, but sometimes read her son's children's Bible with him. Susan reported almost daily readings of *Leading Little Ones to God*, a devotional book, with her children, but neither she nor Janice reported Bible-reading as among their media activities. This was interesting, because both were devoted Bible-readers, and suggests that they understood, as we had in our initial design of the project, that the definition of mass media we were working with at the beginning of the project did not include religious materials.

In their home viewing, the conflicts between class identifications and religious interpretations are the strongest. Most women felt an obligation to censor media, while at the same time being dependent on TV in particular for entertainment, and very familiar with a wide range of shows. Thus mass-market videos and network television programmes and commercials are a big part of everyday media consumption—as are religious reading and church attendance.

According to their week-long media diaries, most of the fundamentalist providers favoured domestic comedies for their television viewing, with *The Nanny*, *Full House*, and *Home Improvement* being among their selections. Most of these, they watched with the entire family. When children were not present, the adults gravitated towards more 'serious' content, mostly dramas, ranging from Missy's favourite choice of *Matlock* to Beth's selection of *Dr Quinn, Medicine Woman*. In fact, Missy noted that it was not the drama of *Matlock* that drew her, but the idea that the ending would be happy and the appeal of Andy Griffith, whom she associated with the clean and simple nostalgia of Mayberry:

> I was just thinking about that [why she watched *Matlock*] the other day. Someone on there's usually getting hurt or something like that, and I thought, usually I don't like shows like that, but there always seems like there's such a happy ending. He always finds them, you know. But if it didn't have a happy ending, I probably wouldn't watch it.

Among the fundamentalist providers we interviewed, Janice Root's home TV viewing habits—and those of her family members—contained the greatest degree of tension between class identifications and religious expectations. For example, Janice said that she and her husband did not allow soap opera viewing in their house because of their immoral content, and then also disallowed *The Simpsons* because of 'Bart going around and cussing and just being a horrible little kid'. Because it was a cartoon, Janice felt that *The Simpsons* implied it was for children, and therefore should not contain rough talk and disrespectful actions. On the other hand, the family watched *Roseanne* together regularly for entertainment:

> Because that, to us, even though it may not be a really good show to watch, it's real life, most of it. I mean, there's some goofy stuff in there, but she's probably the most real-life family that there is on TV, so I don't have a problem watching that . . . I guess we just don't get into real serious TV, you know, real trauma, although we like *[Rescue] 911*.

Rescue 911 appealed to Janice because she liked to see the gripping drama resolved with a safe outcome, and she appreciated its capacity to educate her family on safety issues. All Janice's television favourites are based on working-class characters and / or rural themes.

Husbands were reported to take on some responsibility for deciding on acceptable viewing contents for the entire family in fundamentalist homes. This may have been, in part, because these families were mostly working class. As a related matter, in the Taylor and Root families, men were in the home often; Peter Taylor worked in his home, and Janice Root's husband, at the time of our interview, was laid off from work and performing the majority of the housework. Janice said her husband liked to see the family watch nature documentaries more than other content on television, but that she and the children found these tiresome and successfully lobbied to watch many hours of 'entertainment' television each week. 'You get tired of getting educated', she explained.

By contrast, Susan Gregg felt that she was 'very picky' about what her family watched on television and expressed disappointment that her husband, if left in charge of the children, often did not notice what they were watching. In general, she said, he agreed with her that the children's viewing should be conservatively regulated. Susan alluded in her interview to her disapproval of the more lax TV habits of her co-worker at the school. Susan was conscious of middle-class liberal values, as well as being sensitive to the prohibitions of fundamentalist patriarchy, at times exerting one, then the other set of values. Susan wants her children to trick-or-treat because that is what her neighbours do, but she also criticizes Janice for defying her church's prohibitions on Halloween. Susan took her children to see *Jurassic Park* because it was a popular movie and their friends were seeing it, yet she fears *Barney* and *Power Rangers*. Upward mobility tugs at Susan, as does the fundamentalist sensibility that originates from a working-class position.

Censure from authority figures within the family seemed to be a problem for some women. Beth Mueller spoke with concern about her husband's increasing interest in monitoring their children's television viewing. Missy Jones expressed anxiety about her mother-in-law's disapproval of her videotape collection; the mother-in-law preferred Disney classics and authentic, if 'dark', fairy tales. Janice Root spoke of her embarrassment when her pious mother-in-law criticized her reading of romance novels and her viewing of

Roseanne. Although she eventually gave up the romance novels, her family continued to watch *Roseanne*:

> True, they're [the Conners] not a real religious family, but they still have everyday problems like you have. Even religious people have everyday problems with their kids and with their neighbors and with their sisters, and those are just real things. And they handle things just real funny. And I guess what I like maybe especially about *Roseanne* is that, you know, in all the shows in the past, years and years ago, everything comes to a nice little conclusion in a half-hour and everything worked out for the best—and it doesn't on *Roseanne* and some other shows. But it's still funny to watch.

Laughing, she added, almost nervously: 'I wish they'd go to church.'

Missy Jones also spoke of her immense enjoyment of *Roseanne* until the reference to a women's bar and lesbian relationships placed the programme beyond her frame of acceptability—a reaction we suspect was produced by the explicit demonization of sexuality in her church's doctrine. The exchange bears the mark of her embarrassment, so notable in the conversation about Barbies at school, in discussing sexual content (masturbation, lesbian relations) and also, possibly, her hesitancy in disclosing this position in the interview, until granted 'permission' by the interviewer's question:

> At first, when *Roseanne* first come out [*sic*], we used to watch that a lot. Now, we've completely shut that off. It's just getting too—I think their last episode was about their son, the last one when we finally shut it off, it was about their son, he was in the bathroom . . . [masturbating] . . . And you know, you hear things on the news and things like that, so we finally shut that show off.

At this point Missy is referring to the news story about a *Roseanne* episode set in a women's bar, but seems unable to even refer to it directly. It is left with the vague allusion 'you hear things on the news' until we follow up with a question:

> KAREN: I think the last thing on the news was that she went to a gay bar and they're not going to show the episode?
>
> MISSY: Yeah, 'cause he seen her kissing a woman or something. I thought, Oh, Lordy day! It's not a good show anymore. But when it first come out I thought, you know, it's kind of funny. She was a little bit, still, she'd say things to her kids that I wouldn't probably say to mine, but in a way, I could relate to it because of the way my mother kind of talked to me. She always said things like that to me.

Missy is keenly aware that television's depiction of lesbian and gay relationships are sternly prohibited by contemporary fundamentalists, and it is possible that discussions of *Roseanne* initiated by Dobson and others directly influenced her to change her viewing patterns. Clearly, the media are an important source of entertainment as well as a potential danger—of censure on the job, from relatives, from fellow church members—if these women do not seem to be enforcing the kinds of respectable censorship expected of them because of their role as mothers and teachers. But television is also an important source of entertainment and of other kinds of identification—those of class and gender, for example, rather than religious affiliation. Missy, Janice, and others clearly strain at times against the prohibitions and the irreconcilability of these shifting identifications, perfectly encapsulated in the desire to see Roseanne go to church.

Moral Censure and Non-Christian 'Others'

We sometimes had the impression that these providers embraced what they understood to be a middle-class ideal of respectable Christian rhetoric—one which emphasizes the equality of all in the eyes of God and therefore suppresses the marking of differences. We sensed that this rhetoric may have differed from their own, less respectable private views. It seemed to us that these working-class fundamentalists often navigated their speech with tacit awareness of middle-class Christian authority. Although we asked about racial differences in children's reactions to the media in all our interviews, most of our subjects (and those at the secular schools, who tended to have more education) dodged the question, claiming that they rarely had any children of colour in their care, or simply stating that there were no differences.

It is striking that only once in seventy-five hours of interviews did a teacher mention an incident of explicit racism at the school. During observations at pre-school classrooms and playgrounds in the same town, we have found white children's tendency to mark racial differences and to exclude playmates —especially Asian ones—on the basis of race to be a rather common occurrence. It is perhaps because Missy Jones was the least 'professionalized' of those we interviewed and in many ways the least guarded in her remarks, that she ventured into the following example of a student's undesirable behaviour as part of a discussion about whether exposure to television in a non-Christian home produces bad behaviour:

> Well, matter of fact, one day, at a certain hour, we go out in the middle [a large hall in the middle of the basement pre-school] and play games, and we put all the children together. And a boy in my class told me that he was not gonna play with this black girl [from another class] . . . He said, I'm not gonna play with that brown mama, or something.

Missy concluded her statement with an embarrassed laugh. She recognized the boy's use of the phrase 'brown mama' as unacceptable, and even reported correcting him for it, but she found humour in it herself.

In another example, Janice Root seemed to us to feel an obvious Christian mission in her role as teacher; she spoke with tenderness about the children in her class, but her answer to our question about the children's ethnicities bore the markings of working-class speech: 'They're all white—we don't have any little coloured kids or . . . now I did have a boy for a while who was Asian or Korean or something, I'm not sure exactly.' Popular television shows with African-American characters were absent from the media diaries of all the fundamentalists.

Beth told us about her most challenging case, a boy named Tyrone, whom she took care of for about two years. While most of Beth's complaints sound very similar to those of other care-givers, she devotes a large portion of the interview to voicing her frustrations with Tyrone, whose mother is white and father is African-American—although she does not mention race in her comments. She expressed frustration that, although she tried carefully to regulate her own sons' television viewing, Tyrone's parents allowed him to watch whatever he wanted, including *Teenage Mutant Ninja Turtles*. She said that when Tyrone would start to play violently, using the Ninja Turtles as a model, she had to be careful in correcting him:

> I would try to explain a little bit about why we don't watch them. I would say they were pretend and we don't need to do this stuff. I tried to stress it was pretend. It's OK to play in a fun way, but if it got to be where they were kicking each other, then that was the end of it . . . It was hard for him, because he was used to doing pretty much what he wants at home. He would get upset sometimes. It's kind of hard for me sometimes to know how to say, what he does at home isn't good. I tried to explain we don't play that way.

There is a delicate manoeuvring going on here in her distinction between 'we' and 'you'.

Beth said that what she considered to be the lax care of Tyrone's parents affected her own home. Some of the daycare literature suggests that care-givers such as Beth, who work in their homes, in fact, had many more such problems than people in organizational settings, where rules tended to be made more explicit (Nelson 1990). Peter Taylor, like Beth, complained to us about a boy whose parents were too 'liberal' in what they allowed him to watch on television, and who, as an apparent result, played disruptively. In the instance of Beth and Tyrone, Beth told us that on a few occasions Tyrone's parents had brought him to her home for babysitting in a Ninja Turtles outfit: 'It was too disruptive. It had a mask and sticks that you swing. Everybody would want to put it on. There would be conflict over this. I would make him change, and I would put it up.' Beth described a conversation with Tyrone's mother that clearly echoes a conservative Christian critique of permissive parental behaviour linked to adverse media effects:

> Can you believe [she] was telling me he wanted it in the store and she said no? It was ten dollars. She said, 'He ripped it open, and I had to buy it.' Now I would have just said, 'Too bad, we're putting it back.' You know what they got him for his birthday? He's four years old. A real TV. She told me, 'With Morgan and I both being in school, we just can't give him the attention.' So they put it in his bedroom to watch. I really didn't feel I should be horrified, although I was, and I didn't have much to say.

Here the tension between a non-judgemental professionalized viewpoint on childcare and the conservative Christian's sense of the wrongs of child-rearing in contemporary society come into sharp contrast: 'I really didn't feel I should be horrified, although I was'. She felt secure that her listener will agree with her sense of outrage when she begins with the question: 'Can you believe she was telling me?'

Going to church was linked to monitoring media consumption at home: two important aspects of respectability. These women felt somewhat stifled by religious restrictions on their media consumption, but these same restrictions helped them to define themselves in relation to others. The 'other' for these women is a mother who does not watch what her kids watch on TV: this kind of line-drawing was present in nearly every interview we conducted. But whereas it is used primarily to explain educational failure and behavioural problems, or to define good and bad mothering in interviews with secular teachers and with non-fundamentalist mothers, here it is used to distinguish between Christian and non-Christian families. All the care-givers frequently mentioned, with mild disapproval, the actions of disobedient children of 'non-Christians'.

Scholars studying the sociology of religion have found that fundamentalist parents encourage obedience in their children to greater degrees than do other Christian parents. These scholars have linked the conservative Protestant valuation of obedience with biblical literalism, belief that human nature is sinful, and punitive attitudes towards sinners (Ellison and Sherkat 1993). In other words, for conservative Protestants, the authority of scripture, the fallibility of humans, and the acknowledgement of the necessary punishment of evil-doing combine to make parents fearful of the consequences of their children's sin. They see themselves as having a healthy respect for moral law, and see themselves as enforcers of that law. These beliefs lead them to two kinds of disapproval with regard to television viewing: the permissive television viewing habits of non-religious families represent a lack of discipline and even parental negligence, while the portrayal of permissive parent / child relationships on television, especially situation comedies, is often used to explain the lack of regard for childhood obedience that pervades the culture.

We noticed among the fundamentalist care-providers a blurring between their roles as parents and as care-givers. In the cases of Beth Mueller and Peter Taylor, this overlap of categories was made especially salient because they provided childcare in their homes. Mueller had seriously considered home-schooling her two boys (who were eventually sent to public school); Taylor took his children out of a private Christian school to begin home-schooling them. For both of them, home daycare and schooling are appealing because they offer a way to guard against unwanted influences over their children. But when financial need necessitates taking in other people's children for pay, it means that, once again, unwanted, more permissive, and more media-saturated influences are introduced into the home. It is when conflicts arise between home care-givers and the parents of those in their care that the fundamentalists are most likely to voice their moral disapproval.

When Taylor and Mueller discussed the misbehaviour of the children they take in, they did so with their own family rules in mind. In such discussions discourse about non-Christians may overlap with class, race, or 'lifestyle' prejudices. As in the case of Tyrone's parents, the fundamentalist care-givers in our study were quick to attribute the child's behaviour to the failings of the parents in terms of discipline or lifestyle. While some of this occurred in all of our interviews, the fundamentalist care-givers were quicker to lay blame and rarely extended sympathy to the parents in their struggles to juggle childcare and work. In almost every case, care-givers told us how they would have done things differently from a child's own parent, usually the mother. Such stories were plentiful in our study and almost always related to aggressive boys. Missy Jones, for instance, complained that mothers brought their children to school without matching socks or that fathers brought their children in for care on days off instead of 'spending time with them'.

These fundamentalist care-givers seemed to perceive themselves as surrogate parents for the pre-schoolers in their care—in this their role with the children resembled that of Gloria Williams, the family daycare-provider discussed in the last chapter. They sometimes saw their own influence as especially important when the children are living with parents of whom they do not approve. As Elaine Lawless (1988) has pointed out, maternal authority is one of the few socially acceptable means of power and influence among fundamentalists who must embrace a doctrine of wifely submission. For example, Susan Gregg told us:

> [T]here's so many kids whose lives you're involved in. And there are several kids at school whose home lives are not that great, or who have lifestyles that are not ideal. And it's really neat to be able to . . . be a part of their lives. Try and help them have a little bit of self-esteem. Try and tell them, 'No, you're not bad. You're good! You are just having trouble doing the right thing, you know.' This one little boy—I told him I don't know how many times—and he doesn't seem to ever believe me or he always gets this surprised look on his face when I tell him, 'You're not bad.'

Missy Jones said she wanted to 'protect' the children in her care, comparing their rough home circumstances to her own unsettled family background:

> I feel like I can understand these kids in a way, and I definitely wouldn't want my kids to go through what I went through. That's why I want to love these little kids that are here so much because I understand a lot of them don't have perfect homes. I just want them to know that it can all turn out OK if you . . . help yourself a little bit.

Unlike the general rhetoric of creating self-esteem that permeates the early childhood education literature, and which is explicitly attacked by fundamentalist school-board activists on the grounds that it exalts rather than humbles the self before God (Miner 1996), Missy's way of helping children is to teach her students to pray to Jesus to help them work through problems and to know that Jesus loves them. Although she was at times disapproving of parents, Missy also extended her sense of Christian charity to the mothers. She seems to empathize with other employed mothers, remembering her own feelings about having her children in care while she worked at the university food service:

> I feel like . . . I'm kind of helping these mothers in a way. They probably have never even thought about this, but I feel like, OK, if this was my child sitting here, I would want someone to love it, and I'd want someone to take the time to tie its shoe or wipe its face or hands, and I just feel like that's what I'm trying to do here . . . I'm not their mother and I wouldn't try to replace her, but while they're here, I want them to know someone does love them and is here to be their friend and help them if they can.

Peter Taylor, as the sole male in our sample, shared a sense of guardianship and protection in his roles as parent and care-giver—as suggested by the name of his home daycare centre: 'Little Lambs'. Peter said his fellow church members spoke approvingly about the significance of what he's doing, and he played up the importance of providing a male role model, a theme popularized by the fundamentalist men's organization the 'Promise Keepers', among others. Peter differed significantly from the women we interviewed in his perceptions of the gendering of children's play. For instance, when he talked about television-related play, he focused almost exclusively on male characters, such as Ninja Turtles, and did not mention characters such as Jasmine or Beauty, which are linked with girls. Peter differed significantly from his male counterpart in our earlier sample, Glenn McClintock, however, in that Glenn was extremely offended by Barbie for what he saw as its encouragement of narcissism in the girls and traditional gender schemes—he felt that the school should be much stricter about banning Barbie from show and tell. Peter's discussion of Disney movies showed an absence of the gender-role critique found in the remarks of some of the fundamentalist women—a concern, for example, that the male characters were too violent.

Like Glenn McClintock, Peter did not show the degree of concern that the women did about boys' rough play involving television characters. As he explained, 'I tend to be less afraid of rougher play because of memories of doing it . . . If it's going to be detrimental to someone's health, I would step in.' Glenn's comments about boys' rough play are almost identical to Peter's. Although Glenn identified himself as a pacifist who was strongly opposed to violence in the media (just as Peter is especially focused on Satanism and supernaturalism), he felt that the women teachers were frequently overly alarmed by and improperly understood rough-housing. This suggests that there may be important gender differences in adult attitudes towards children's play and the need for supervision and restriction. Investigating the ways that these attitudes correlate with experience in childcare as well as with express religious or political ideologies would be an important contribution to the literature on children and media. An important question in this regard would be whether men in their roles as fathers and childcare-givers are less influenced by the proscriptions of experts—religious or secular—in their practices than are women, for whom the acceptance of various forms of authority is a large measure of the experience of motherhood (Seiter 1996).

Conclusion

In closing this chapter, we would like to examine some of our motivations for studying fundamentalist women, and some of the challenges that such pilot research suggests for a full-blown ethnography of Christian fundamentalists and the challenges they pose for the study of television in everyday life.

There is doubtless a kind of voyeuristic as well as a politically motivated interest in representing the lives of these women. While this was not openly discussed in the interviews, the women themselves were almost certainly aware of the risks involved in talking to university types. As suggested by folklorist Elaine Lawless's work with evangelicals, these women probably came to the interview with some suspicions and distrust of the possibility that they might come off as 'crazies':

> Pentecostals are clearly aware of the stereotypes that non-Pentecostal people hold about them. They are aware that people view them as 'freaks' and refer to them in a derogatory way . . . [They] wish both to maintain the important boundaries between themselves and 'the outside world' *and* to illustrate that they actually are not *all* that different from other people. (1988: 59)

The fundamentalist pre-schools did screen our requests carefully, and some of our respondents, especially the home-schoolers, were unusually difficult to track down for the follow-up interviews. Although our sample was small, the circumstances of these interviews—a research project on children's TV watching rather than an investigation of the women themselves—may have provided an element of security and thus an unusual opportunity to profile some of the positions of fundamentalist women without making them feel defensive. In their discussions of media and consumer culture, they were both doctrinaire (as when they unveiled incantations and satanic themes everywhere in children's media) in ways that suggest the powerful influence of groups such as Focus on the Family, and unconstrained by dominant interpretations (wishing to defend *Roseanne*; seeing Barbie as a wholesome, church-going mother).

From our side, we recognize that this work seems to fit rather well into one of the typologies of feminist cultural criticism proposed by Charlotte Brunsdon, namely the hegemonic phase characterized by an interest in 'non-feminist women others'. This category is described by Brunsdon:

> The construction of feminist identity through this relation involves the differentiation of the feminist from her other, the ordinary woman, the housewife, the woman she might have become, but at the same time a compulsive engagement with this figure. The position is often profoundly contradictory, involving both the repudiation and defence of traditional femininity. (1993: 313)

Our relation to these subjects involved both fear (of their politics, of their worldview) as well as familiarity (with their struggles to care for small children, their desires to seek out both entertainment and sociability in their leisure time). One of us, coming from an urban, Northern, middle-class, and Catholic background, felt a much greater degree of fear, however, while the other found a much greater degree of familiarity, coming from a working-class, Southern background, where some family members practised 'old-time religion'. As Linda Kintz and Julia Lesage (forthcoming) have pointed out, these kinds of differences, of region and cosmopolitanism, are precisely those that have been so effectively exploited by the Christian Coalition in organizing white women like the ones we interviewed.

As feminist intellectuals explore the terrain of Christian fundamentalism and its female adherents as a kind of absolute other, it may do well to bear in mind Karen McCarthy Brown's warning about what she describes as 'the crossfire of projections that occur when fundamentalists and scholars describe one another':

> Scholars share with many fundamentalists a tendency to invest texts with great authority . . . Both groups engage in the pretense of being fully in control of what we deal with. Scholars and fundamentalists both operate with a view of self that stresses consciousness and discipline, while denying that deep hungers, fears, and needs influence our search for meaning. We also have other, more subtle matters in common; we are both hierarchically organized; we jealously guard our boundaries; we quibble over whom we will admit to membership; and we make neophytes go through more or less elaborate initiation rites. Most important, scholars and fundamentalists alike are currently threatened by the potential breakdown of their hegemonic world views. (K. M. Brown 1994: 196)

As working-class women in American society face mounting challenges—ranging from growing economic adversity to the erosion of political rights—the predicament of women who have allied themselves with fundamentalism becomes more important for scholars to confront. We believe it is essential not to overlook the tensions and contradictions in these women's lives. We should neither gloss their rhetoric as exclusively patriarchal, nor mistake it for feminist resistance. The fundamentalists we found here, especially the women, were struggling to exert their influence in the one sphere where they felt powerful—child-rearing. Our study suggests the importance of understanding the material benefits offered by religious involvement to women at this precise historical moment—a trend with global, as well as national, significance. As Helen Hardacre notes, 'In many regions of the developing world, fundamentalist movements provide their own social programs in the absence or inefficiency of programs provided by the secular state' (1993: 137). Certainly, the government's failure to provide adequate childcare, or to attend to the needs of families with small children, is one of the factors exploited by the Christian Right in its appeal to women.

This appeal merits closer and more serious attention from feminists and members of the left who too often dismiss fundamentalists as ridiculous victims of false-consciousness. An important question for future research will be how it is that fundamentalist and Pentecostal churches attract women by simultaneously taking their problems of child-rearing and moral authority over children seriously, while at the same time expressly advocating ideas about gender that are explicitly at odds with the feminist agenda.

Media scholars should also note the growing number of conservative Christian media companies which are eager to supply educational and entertainment products (often through marketing on the World Wide Web) to the rapidly increasing numbers of parents-turned-teachers in the Christian home-schooling movement. All the teachers in this pilot study used prepackaged curricular materials, something that facilitated their ability to take on the job without any specialized training (and thus contributes to the deskilling of the labour of early childhood education). The value of such materials stems in part from the certainty that these products will follow a strict line of parental authority, sexual chastity, creationism, and Bible education. In the current climate of narrow casting and niche marketing, will Christian media companies target fundamentalists as an alternative market to that of network television and mainstream consumer culture? In this study we have treated religion as quite separate from our subjects' pursuit of entertainment: would it be more accurate to conceptualize religious media consumption and church activities along the lines of participation in popular culture and leisure entertainment?

The fundamentalist critique of the mass media parallels as well as deviates from the secular scares about mass media effects (Gilbert 1978) and the pronouncements of childhood experts such as psychologists and educators. With renewed concern about the increase of violence on television, the lack of content regulation on cable, and the weakening of Standards and Practices Departments, strands of the political right and left appear to be coming together in their advocacy of renewed censorship. As historian James Gilbert usefully reminds us, the concern over the mass media's corrupting influence on children, its weakening of parental authority, and its ability to inculcate values, is an 'episodic notion': one that could be introduced at any time and thus must be studied for its social function, and for how it inevitably reflects and shapes social relationships. The problem with the old worry about mass culture's effects on children is that 'the control of youth is, even in the best of times, problematic, and . . . the cultural history of the United States is filled with dramatic changes in the content and form of popular culture' (Gilbert 1978: 4). Thus, the timing of public concern over media's effects on youth and the ways it is introduced into secular, political, and religious discourses warrant careful scrutiny.

More research on the parallel trends of expert opinion and religious opinion regarding authority over children and authority over the media is needed, research that would attend to the negotiation and blurring of these categories under specific historical circumstances. The childcare-givers in our study were moving back and forth between a more mainstream middle-class critique of mass media (as too violent, too commercial, too precocious) and a fundamentalist one (too secular, too sinful, and too satanic). Both of these may be very traditional in American culture, as Gilbert's work suggests, and very embedded in many streams of formal and informal thought. This topic deserves study with a sample more diverse and extensive in terms of region, class, religious affiliation, race, and ethnicity. Our work suggests that in-depth qualitative methods, such as an ethnography of a church community, could offer a rich picture of the ambiguities and contradictions lived by many Americans today through their beliefs and practices about gender, about religion, and about the relationship of adults to children. The problem, for many of them, is how to negotiate a comfortable path between the traditional values they embrace and the media and consumer culture that surround, challenge, offend, and often attract them.

6

Television and the Internet

CLAIMS THAT THE Internet will revolutionize communications (as well as education, work life, and domestic leisure) are now commonplace. Yet there is a danger that computer communications—and by this I mean their uses, the discourses surrounding computers and the Internet, and research about them—will substantially buttress hierarchies of class, race, and gender. One healthy corrective is to recognize the many parallels between television and the Internet, and incorporate the insights of television audience research into the uses of technologies in the domestic sphere, the articulation of gender identities through popular genres, the complexity of individuals' motivations to seek out media, and the variety of possible interpretations of media technologies and media forms. Ethnography can offer a rich context of understanding the motivations and disincentives to using computers: an important research topic in a world in which non-users are likely to be labelled recalcitrants, technophobes, or slackers.

Most academic research on digital technologies is currently being produced by departments of library and information science, schools of business and management, education schools, and computer science departments. This research tends to emphasize information-seeking and statistical patterns of usage, while ignoring perceptions about computers, the cultural contexts in which they are used, and the images, sounds, and words to be found on computer screens. We need a means of touching upon the form and content of the Internet as well as the practices and motivations of computer users. We need to view the Internet from the perspective of its many parallels with broadcast media, maintaining a healthy scepticism about its novel qualities as a communication medium. We need to be alert to the ways in which stereotyped notions of the audience are constructing a discourse around the Internet that privileges white, middle-class males.

In this chapter, I begin by reviewing the ways in which familiar forms of television are migrating to computer screens, while television, for its part, is busily promoting use of the Internet by television viewers. These connections between television and computers are taking place at the level of corporations, as Microsoft attempts to enter the mass-market entertainment business by investing in media firms; at the level of technology, as computers with television tuners and video stream capabilities become more commonplace; and at the level of form and content, as familiar genres from television, radio, and newspapers are tried out on the Internet, many of them sponsored by such giants of television advertising as Procter and Gamble and Nabisco. In the second section, I survey qualitative research on the use of computers in domestic settings among families, and trace the similarities between these findings and research on television viewing in the home. I argue that television audience research is well positioned to help us understand the heavily gendered use of computers in the domestic space. At the same time, research on computer use may help to push television audience research to a more thorough investigation of the connections between domestic and public uses of media, and to think more about television viewers as workers, not only as family members or individual consumers. In the third section, I look at the various work issues related to computer use, the increase of 'telecommuting', and its implications for the study of communication technologies in the domestic sphere.

Television on Computers

Television sets and computer terminals will certainly merge, cohabit, and co-exist in the next century. In 1996 computers with built-in television tuners became available, as did set-top boxes to allow Internet access via television sets. Because of the proliferation of television sets throughout many homes, they are increasingly likely to share space with computers in the same room. Many people (most of my students, it seems) have become adept at watching television while using the computer. As the number of personal computers increases in middle-class homes, the boundaries between leisure and work time, public and private space, promise to become increasingly blurred (Kling 1996). As the Internet develops from a research-oriented tool of elites to a commercial mass medium, resemblances between Web sites and television programming will increase.

The Internet can be used to organize users around political matters in ways unimaginable through broadcast television or small format video—and fundamentalist Christians are one group that has already proven this potential. It seems increasingly likely, however, that commercialization of the Web will discourage activism in favour of consumerism and the duplication of familiar forms of popular mass media, such as magazines, newspapers, and television programmes (Morris and Ogan 1996). The World Wide Web reproduces some popular genres from television (and radio) broadcasting: sports, science fiction, home shopping clubs, news magazines, even cyber-soap operas with daily postings of the serialized lives of its characters. In fact, the most popular Web sites represent the same genres—science fiction, soap operas, and 'talk' shows—that form the topic of some of the best television audience research, by, for example, Press (1991), Jenkins (1992), Gillespie (1995), and Shattuc (1997). The much-publicized presence of pornography on the Internet also parallels the spectacular success of that genre on home video.

The prevalence of television material on the Web confirms the insight provided by media ethnographers of the importance of conversation about television in everyday life, and suggests that television plays a central role as common currency, a *lingua franca*. Television fans are a formidable presence on the Internet: in chatrooms, where fans can discuss their favourite programmes or television stars; on Web sites, where fan fiction can be posted; and as the presumed market for sales of television tie-in merchandise. The dissemination of knowledge of the programming language for the creation of Web sites (or home pages) unleashed countless die-hard television fans eager to display their television knowledge—and provide free publicity for television producers. Hundreds of painstakingly crafted home pages have been devoted to old television shows. For example, one site devoted to the 1964–5 Hanna-Barbera cartoon *Jonny Quest*, provides plot summaries and still frames of every episode ever made. In fact, the Web is a jamboree of television material, with thousands of official and unofficial sites constituting television publicity, histories, cable and broadcast schedules, and promotional contests. Search engines turn up roughly three times as many references to television as they do for topics such as architecture, chemistry, or feminism. Apparently, television was one of the first topics people turned to when trying to think of something to interest a large and anonymous group of potential readers—other Internet users.

It would be a mistake, however, to see the rise of television material on the Internet exclusively from a fan or amateur perspective, because the connections between television and computer firms are proliferating. The association between television and the Internet has been heavily promoted at the corporate level by access providers eager to lure as sponsors companies that invest heavily in television advertising, and by others seeking sources and inspiration for the new Internet 'programming' (Schiller 1997). The software giant Microsoft corporation has acquired stakes in media entertainment companies, developed an interactive television network, and looked to television and film for the basis for entertainment 'software' with a more 'universal', that is, mass-market, appeal. Microsoft's partnership with NBC to form the twenty-four-hour news cable network and on-line news and information service MSNBC is the most obvious example. Microsoft also has joint-venture deals with the cable network Black Entertainment Television; with Stephen Spielberg, Jeffrey Katzenberg, and David Geffen's company Dream Works SKG; and with Paramount Television Group—all of which is leading to speculation that Microsoft is 'morphing into a media company for the new millennium' (Caruso 1996). Disney, now the owner of the television network ABC, is also one of the biggest interactive media producers in the world. America On-Line, the commercial Internet and e-mail access provider, has followed a vigorous commercial strategy, which includes extensive coverage of television in all its familiar publicity aspects as well as encouragement of fan activity, to build a broad base of subscribers and to court advertisers. The A. C. Nielsen company, the television industry leader in audience ratings, produces reports on Internet users. Worldgate Communications is offering a set-top box and hand-held remote control device that allows a viewer to access Web sites tied to television programmes currently being watched.

In 1996, Microsoft made its intention to break into broadcasting explicit when it released a revised version of its on-line service, whose browser interface sends users straight to an 'On Stage' section with six different channels, each hosting 'shows' (Helm 1996). The goal was to give users a better idea of what to expect from each programme by standardizing its offerings, a strategy strikingly familiar from the history of early radio and television (Boddy 1990; McChesney 1996). Video-streaming, already a commonplace on the World Wide Web, has been implemented on Web sites such as Cable News

Network's to replay the 'News Story of the Week'. Advertising industry analysts predict that animated advertisements on the Web will dominate in the years ahead.

Hardware and software manufacturers are scrambling to secure the market for sales to non-computer owners of devices that will convert the plain old television set to an Internet browser, or win the battle between the high-definition TV sets favoured by the electronics industry and the digital television / monitors favoured by the computer industry. The computer position is that 'consumers would rather have a cheaper box that would be either a computer monitor or a TV than have the less complicated, high-definition-TV set that the consumer electronics industry favors' (Auletta 1997: 77). Microsoft is exerting considerable muscle in political lobbying and industry influence over issues of digital television (Bank and Takahashi 1997), exemplified by Bill Gates's decision to purchase WEB TV for $425 million and to announce this decision at the annual convention of the National Association of Broadcasters.

The relatively high penetration rates of home computers among the professional classes (including writers about computer issues) often give the false impression that everyone has a computer. The majority of homes do not have a PC, but they do have a television. Therefore, the computer industry continues to eye television greedily as a future market. Microsoft has entered the cable television business, exploring set-top boxes and television programming, and vigorously campaigning to thwart the success of High Definition Television (HDTV) in favour of 'digital TV', which would use computer monitors. Television set manufacturers are gambling on a variety of designs that integrate Web access with television viewing through cable boxes, double windows, wireless keyboards, and television remote controls with data entry features. Digital TV sets are being designed to maximize flexibility for future uses with satellite receivers, Internet navigators, digital video disc-players, and set-top boxes.

The extent to which Microsoft is explicitly using television and its mass appeal as a model for future endeavours was made explicit in a recent article by Ken Auletta about Nathan Myhrvold of Microsoft. Microsoft realizes that 'the skills that made Microsoft successful in software—technical proficiency, rapid response—are not transferable to what the company calls the content business, which relies more on a bottom-up rather than a top-down model' (Auletta 1997: 76). The fact that personal computers are stuck at a penetration rate of about 36 per cent has led computer industry people to eye television jealously, and to try to develop ways to link perceptions about computers to entertainment. At Interval Research Corporation, the think-tank started by Microsoft's co-founder Paul Allen, the Explorers market research group has adopted the strategy of using television—a universally accepted domestic technology—as a model for the development of future communication technologies (Ireland and Johnson 1995).

For its part, television plays a crucial role as publicizer of the Web and computer use. Television programmes are already filled with references to computers and the Internet that both dramatize the importance of the new technologies and attempt to play a major role in educating the public about new media. Television's appetite for novelty, as well as its fears about losing viewers to computer screens, make computers one of its predictable obsessions. Silverstone and Hirsch are right to point to the dual nature of communication technologies such as television sets and computers 'as quintessentially novel objects, and therefore as the embodiment of our desires for the new', which simultaneously act as 'transmitters of all the images and information that fuel those desires' (1992: 3).

Computer references have gone far beyond the television character staring into the computer screen—although this long-standing movie cliché has been solidly established as a convention of television drama. Television commercials refer viewers to Web sites; call-in programmes now ask that e-mail be submitted, as do television shows from *Meet the Press* to Nickelodeon's children's line-up. Numerous cable shows include e-mails (reflecting the striking banality of much chatroom conversation) scrolling across the screen below the programme material. Television news shows, especially docudramas and news magazines, have become so enamoured of reproducing e-mail and Internet communications (which are both easy to capture on camera and lend a feeling of novelty and a sense of connection to the real world of viewers), that the practice has become a copyright / privacy concern among computer specialists (Lesch 1994).

On-line communications have been used both to support and to attack television shows and their sponsors. Television networks are exploiting e-mail and Internet communications with audiences to gain feedback on script or character changes, to compile mailing lists for licensed products relating to shows, and to publicize tie-in merchandise. The creation of an Internet home page for the *X-Files* was credited with saving the show from cancellation after its first season. The *X-Files* producers recognized the perfect synergy of its high demographic fans and the Internet, and targeted its audience through the World Wide Web, a move that helped both to prove its audience share to executives contemplating axing the show, and to generate more publicity for the programme. Protests against television animate the on-line communications as well: Christian Right organizations such as The American Family Association use the Internet to organize protests against television sponsors of objectionable material (a list it calls 'The Dirty Dozen') and 'filth' (*NYPD Blue*).

Computer and Internet research can benefit greatly from television research on television flow and the use of remote controls, the instalment of the television as a domestic object, and conversation around television. For example, the Internet poses problems similar to that of television 'flow' (Williams 1974), as Web 'programmers' (especially those with commercial sponsors) attempt to guide the user through a pre-planned sequence of screens and links. While nearly every branch of the advertising industry is making moves to work on the Internet, anxieties are already rife about ways to measure consumers. At first, the number of 'hits' a Web site received was enough to entice sponsors. Repeated anxieties have surfaced about the fact that banner advertisements are often ignored and software that attempts to track the behaviour of Web users cannot detect when the transfer is stopped before the ad is delivered. Thus, talk of hits gave way to a preference for 'impressions'—a word more likely to carry weight with sponsors, with its desirable associations with lasting mental influence. While an estimated one billion dollars was spent on Internet advertising in 1997, and that figure is expected to double in 1998, advertisers remain sceptical about tracking software and accountability in measuring consumers (Matzer 1998). This concern for the meaningfulness of exposure to Web advertising closely parallels the anxieties of television advertisers about the attention span of television viewers, and their proclivity for 'zapping' commercials by switching channels, or 'zipping' past commercials on videotapes of pre-recorded programmes. From the advertiser's perspective, Web surfers can be just as fickle as television watchers, it seems.

In *Desperately Seeking the Audience* (1991), Ien Ang carefully deconstructed the fantasies of control over television viewers and the necessity of such fantasies to the daily functioning of the television industry. In the trade publication *Advertising Age*, these fantasies, and the battle over competing claims for accurate measurement of Web surfers, are now a major preoccupation. On the one hand, the Internet is projected as a much better vehicle than television, because the Web user is persumably more attentive, more goal-directed—and wealthier. Yet, on the other hand, anxieties about measuring and controlling Web users are escalating, and energies are focusing around the development of measurement devices adequate to convince sponsors.

For its target market of the professional upper-middle class, advertisers are promoting the installation of the computer as a domestic object in a process similar to the guidelines for installing the television set in the home in the 1950s studied by Lynn Spigel (1992). Computers are advertised on utopian claims to enrich family life, enhance communications, strengthen friendship and kin networks, and, perhaps most importantly, make children smarter and give them a competitive advantage in the educational sphere. In advertising, in news broadcasts, in education journals, the computer is often defined against, and pitched as an improvement on the television set: where television viewing is passive, computer use is interactive; where television programmes are entertaining in a stale, commercialized, violent way, computer software and the Internet are educational, virtuous, new.

In this book, I have stressed the ways that negative feelings about television viewing (shame or defensiveness) affect what people are willing to say about television. Comparisons between television viewing and use of the World Wide Web are inevitable. Like television programming, computer media—software, Web sites, etc.—serve as topics of conversation, but the latter hold more legitimacy among the educated middle classes. Among middle-class professionals, the group best positioned to parlay computer use into improved earning power, discussing a new Web site holds more cachet than talking over last night's sitcom. The negative associations of being a computer nerd, or even a hacker, have abated considerably in the last decade (Turkle 1995), while computer magazines such as *Wired* have promoted fashionable postmodern associations with computer use. While some sanctions are associated with being a nerd, this stereotype has a higher gender, class, and intellectual standing than the couch potato. On the other hand, those with less disposable income and less familiarity with computers may reject computers for the values they represent (such as dehumanization), their emphasis on written rather than oral culture, their associations with

Computers and TVs advertised with utopian claims to enrich family life

white male culture (hackers and hobbyists), and their solitary, anti-social nature. The operations of 'distinction' will be especially important to bear in mind when doing empirical work on the social contexts of computer use.

Gendered Uses of Computers at Home

Television sets and computers introduce highly similar issues in terms of placement in domestic space, conflicts among family members over usage and control, and value in the household budget. We can expect these conflicts to be articulated within gender roles in the family. Some research on gendered conflicts over computers (Haddon 1992; Murdock *et al.* 1992; Giacquinta *et al.* 1993) reproduces themes of family-based studies about control of the television set. Already, researchers have noted a strong tendency for men and boys to have more access to computers in the home. Television studies such as those carried out by Ann Gray (1992), David Morley (1986), and myself (Seiter *et al.* 1989) suggest that women in nuclear families have difficulty watching a favourite television show (because of competition for control of the set from other family members, and because of shouldering the majority of childcare, housework, and cooking). If male

The good screen: computers as purveyors of high art

family members gravitate towards the computer as hobbyists, the load of chores relegated to female family members will only increase, and make it more difficult for female members to get time on the home computer. Computers require hours of trial-and-error experimentation, a kind of extended play demanding excess leisure time. Fully exploring the Internet demands time for lengthy downloading, and patience with connections that are busy, so much so that some have dubbed the World Wide Web the World Wide Wait.

In the family, computers can create anxiety, too: young children must be kept away from the keyboard because of potential damage to the machine. Mothers, who have traditionally been charged with securing the academic success of their children, would have a strong incentive to relinquish computer time to older children, who are thought to benefit greatly from all contact with the technology. When anxieties increase and moral panics are publicized about children's encounters with pornography through the computer, or the unhealthy effects of prolonged computer use, the brunt of responsibility for enforcing restrictions on computer use will fall to mothers and teachers.

Some qualitative research has already explored these areas, and some of the most valuable work has been informed by British cultural studies work.

Silverstone *et al.* have offered a fascinating case study of a well-to-do London family whose home included a wide array of new technologies, and whose explicit ideology was one of encouraging children to use them. Yet the mother remained at a weary distance from the computer. She responded with irritation to researchers' questions about her feelings towards the communication technologies, claiming not to have feelings about technologies at all (1990b: 35). Both parents desired that their children gain computer experience and preferred this to television watching, but expressed irritation with the boy's domination of their home computer. Measures were taken to try to secure computer time for the daughters in the home, but with mixed success. The mother felt alienated from the developing 'father and son' culture around the computer, and suffered arguments about the selection of computer games, but her frustration led her to take a course on computing. Tensions among highly motivated, well-educated females over computer technology deserve much more investigation.

Similarly, Giacquinta *et al.*, in their qualitative study of white middle-class New York families, found marked differences between males and females, and between adults and children, including less use by females: 'mothers were particularly estranged from the machines' (1993: 80). The study, conducted in 1984–7, included sixty-nine mothers, two-thirds of whom were employed full time outside the home. Mothers tended to use computers for word-processing and 'did not engage in programming, tinkering, pirating, or game playing' (1993: 81). They found the violence in computer games objectionable. In general, they lacked 'the interest, the need and the time' to develop computing skills (1993: 89). Daughters were more likely than mothers to use the computer, and had more resources to support them, such as classroom teaching, but they were not a focus of the girls' leisure-time activities.

Friendships, kin networks, and work relationships are crucial to the successful adoption of new technologies such as computers (Douglas 1988). Computer use often involves borrowing software, troubleshooting problems, trying out new programs, boasting or discussing successes, and cross-checking machines. Advice and encouragement are important components of this. The Giacquinta study found women rarely spoke with other women about their computers or assisted each other in learning. If women and girls tend not to talk about computers, they are at a sizeable disadvantage over boys and men, especially those with considerable practice at hobby talk. In another study of home computing, in the English Midlands, Murdock *et al.* found, in a sample of one thousand households, that males outnumbered females as the primary computer user by a ratio of seven to one. They also found that those who did not talk to others about computers or borrow programs from friends or relatives were most likely to have stopped using their computers (1992: 150).

Jane Wheelock found, in a British study of thirty-nine families in 'a peripheral region of the national economy', that there were three times as many sons as daughters interested in the home computer. Wheelock's study focuses on the ways that the household, operating as a complementary economy to the formal one, reproduces and produces labour power as related to computer use (1992: 98–9). Daughters were more likely to be interested in computers if such interest was facilitated by parental and teacher encouragement, or if there were no sons in the family—or after the machine had been abandoned by the sons in the family (1992: 110).

Boys were much more likely than girls to use computers as a part of their social networks, something that, Wheelock notes, 'increases boys' socializing, and shifts its locus towards the home; traditionally both are features of girls' experience' (1992: 111). Lesley Haddon's observations, based on time spent in a computer club, similarly suggest that girls may not use computers as a topic for school conversation even when they do use computers at home, being more likely to discuss rented videos (1992: 91). The girls in Haddon's study were also unlikely to play computers in public places, stores, arcades, or school clubs, but used them at home.

Most of these studies are somewhat dated, and do not provide any information about the use of electronic mail and the World Wide Web. These two uses of the computer, in their facilitation of personal communications, most closely resemble the telephone, a communications technology particularly valued by women (Rakow 1988; Livingstone 1992; Spender 1995). Reliable information about Web users is hard to come by, and most research relies upon self-selection of its sample, that is, people responding to various postings asking for Internet users to fill out a survey form. Some of this recent survey information suggests that the Internet may be attracting women, especially younger, white, middle-class women under 35, at surprising rates. In 1995, the Georgia Tech World Wide Web users survey reported that 29 per cent of their respondents are female—a number that has increased significantly over the last three years. There has been a substantial increase in female Web users between the ages of sixteen and twenty, and an increase in female users who teach kindergarten through twelfth grade—there was a 10 per cent increase in female users in this category observed in one year, 1994–5. The importance of access through public education institutions for women is significant: 39 per cent of women responding gained Internet access that way, compared to 28 per cent of men. Yet there are clear signs that women are less likely to use computers intensively in their free time. At the weekends—an indication of hobbyist users—the gap widens to 75 per cent male and only 25 per cent female: thus, three times as many men are weekend users of computers compared to women. Women users were less likely than men to spend time doing 'fun computing' or to use a computer for more than thirty-one hours a week. Although the Web is attracting men over the age of 46 in increasing numbers—many of them as a retirement hobby—the relative numbers of women of that age who use the Internet are declining.

Much has been written about the ways in which the Internet can be used to explore personality and identity (for example, Turkle 1995; Star 1995). But women and the poor are going to be less advantageously positioned to engage in such activities for a complex set of reasons (Star 1995). As Roger Silverstone has explained:

> [T]he ability to use information and communication technology as a kind of extension of the personality in time and space . . . is also a matter of resources. The number of rooms in a household relative to the number of people, the amount of money that an individual can claim for

Using computers 'as an extension of personality'

his or her own personal use, the amount of control of his or her own time in the often intense atmosphere of family life, all these things are obviously of great relevance. (Silverstone 1991: 12)

Computerized Work

Are women more likely to use the Internet if they use computers on the job? Working on a computer can mean very different things: if we are to understand the differential desire to use computers during leisure time, it is essential to make distinctions between kinds of computerized work. For example, huge numbers of female employees occupy clerical jobs that use computers for processing payroll, word-processing, conducting inventory, sales, and airline reservations—more than 16 million women held such positions in the USA in 1993 (Kling 1996). Women overwhelmingly outnumber men in the kinds of job where telephones and computers are used simultaneously:

airline reservations, catalogue sales, telephone operators. In contrast, fewer than half a million women work as computer programmers or systems analysts.

The type of employment using a computer that is likely to be familiar to the largest number of women, then, is a kind of work where keystrokes might be counted, where supervisors may listen in on phone calls, where productivity is scrutinized on a daily and hourly basis, where conversation with co-workers is forbidden (Clement 1994; Iacono and Kling 1996). The stressful and unpleasant circumstances under which this kind of work is performed might explain women's alienation from computer technology and their tendency to stay away from it during their leisure time.

Some parallels exist between clerical work and teaching—one of the sole white-collar professions dominated by women. As Steven Hodas points out in his discussion of what he calls 'technology refusal' in schools, most teachers are women, and most educational technologists are men, who target their efforts at introducing computer technologies towards classroom teachers, not male administrators. Too often, technologists express the need for classroom computers in ways that derogate the work of teachers: in these discussions, 'the terms used to describe the insufficiency of the classroom and to condescend to the folk-craft of teaching are the same terms used by an androgenized society to derogate women's values and women's work generally'. When technologies fail in the classroom, the reaction is to 'blame the stubborn backwardness of teachers or the inflexibility and insularity of the school culture' (Hodas 1993: 206). Hodas usefully reminds us that present-day arguments about the need for computers in schools mirror the same redemption through technology arguments that accompanied other media, most notably for our purposes, educational video:

> The violence that technologists have done to our only public children's space by reducing it to an 'instructional delivery vehicle' is enormous, and teachers know that. To abstract a narrow and impoverished concept of human sentience from the industrial laboratory and then inflict it on children for the sake of 'efficiency' is a gratuitous, stunning stupidity and teachers know that, too. Many simply prefer not to collaborate with a process they experience as fundamentally disrespectful to kids and teachers alike. (1993: 213)

Telecommuting, working from the home through a modem or Internet access to the office, is a different category of computerized work which is supposed to hold special appeal to women. Telecommuting is now officially sanctioned by the US government, according to Rob Kling:

> A recent report developed under the auspices of the Clinton administration included a key section, 'Promoting Telecommuting,' that lists numerous advantages for telecommuting. These include major improvements in air quality from reduced travel; increased organizational productivity when people can be more alert during working hours; faster commercial transit times when few cars are on the roads; and improvements in quality of worklife. The benefits identified in this report seem so overwhelming that it may appear remarkable that most organizations have not already allowed all of their workers the options to work at home. (1996: 212)

The dream of telecommuting

While there are growing numbers of women doing pink-collar jobs such as clerical and sales work at home—that is, using their home computers at jobs such as credit-card verification and telephone solicitations—this form of telecommuting is also largely invisible in the mass media. Instead, we see images of female professionals using computers to work from home, perhaps while their one and only child conveniently naps in the next room. As Lynn Spigel has noted, computer publications suggest 'a hybrid site of home and work, where it is possible to make tele-deals while sitting in your kitchen' (Spigel 1996: 11). In her discussion of a *Mac Home* magazine cover, Spigel suggests that: 'The computer and Net offer women a way to do two jobs at once—reproduce and produce, be a mother and hold down a high powered job. Even while the home work model of domestic space finds a place for women, it does not really break down the traditional distinctions between male and female' (Spigel 1996: 12). As more workers struggle to get their work done from home without the benefit of an office, questions about the use of communication technologies become especially interesting and well suited to ethnographic approaches.

Another interesting issue for audience research is the entertainment uses of computers in offices, as office workers have access to more entertainment and play functions through their computers. Sherry Turkle's book *Life on the Screen* (1995) examines a group comprised mainly of students or white-collar computer programmers who work on computers throughout the day. Turkle is especially interested in participation in Multi-User Domains, or MUDS: on-line fantasy games that can have dozens of players. Turkle describes this balance of play and work:

> I have noted that committed players often work with computers all day at their regular jobs. As they play on MUDs, they periodically put their characters to sleep, remaining logged on to the game, but pursuing other activities. The MUD keeps running in a buried window. From time to time, they return to the game space. In this way, they break up their day and come to experience their lives as cycling through the real world and a series of virtual ones. (1995: 189)

While Turkle is primarily interested in the psychological dynamics of play with a variety of virtual selves, she rarely foregrounds the very specific class fraction which has the privilege to play on the job. In this world, white-collar, upper-middle-class employers are finding that after a period of vigorous encouragement if not requirement of nearly constant computer and Internet use in many occupations, employees are spending large parts of the day playing computer games, writing personal e-mails, and cruising the Internet, and that it is increasingly difficult to confine white-collar employees to work-related rather than entertainment uses of these technologies—or in some cases to distinguish between the two. As one expert wryly put it in a discussion of the impact of digital video discs (DVDs), 'The big application for DVD later this year will be desktop video playback, which will eliminate any remaining worker productivity that hasn't already been destroyed by Web surfing' (Hood 1997: 14).

Turkle gives many examples of women immersed in the play with identity and the capacity for writing one's own dramatic narratives that MUDs offer, and she is especially interested in the phenomenon of players impersonating the opposite sex in their virtual personae. Turkle's study fails to interrogate the particular class background of the MIT students and computer programmers she interviewed for her study, or to ask how particular class positions may predispose women to be attracted to the kind of play with virtual selves that she describes as postmodern. The luxury of such computer play is unimaginable in the context of most clerical computer work and unlikely in the home workstation when women race to get their work done before children come home from school, or hurry to turn their attention from the computer to housecleaning, shopping, or food preparation.

For writers such as Turkle, the Internet offers an arena for exploring—and deconstructing—traditional gender roles that can be quite liberating. Such an analysis contrasts sharply with the concerns of Cheris Kramarae, who, in her analysis of the genres and kinds of information available on the World Wide Web, has noted the rapid proliferation of genres such as pornography, and modes of discourse, such as flaming, which may act as deterrents to women going on-line (Kramarae and Kramer 1995). Similarly, linguist Susan Herring has cast doubt on utopian claims for on-line communications in the workplace that have suggested these technologies may be advantageous for women getting ahead in professions. This work suggests that even among white-collar professionals, computers may be more fun for men than for women. For example, in a study of academics using electronic discussion groups, Herring found that:

> [M]ale and female academic professionals do not participate equally in academic CMC [Computer Mediated Communication]. Rather, a small male minority dominates the discourse both in terms of amount of talk, and rhetorically, through self-promotional and adversarial strategies. Moreover, when women do attempt to participate on a more equal basis, they risk being actively censored by the reactions of men who either ignore them or attempt to delegitimize their contributions. (1996)

Herring concluded that there was nothing inherently wrong with the technology of computer bulletin boards; rather, the problem stemmed from the ways that old, familiar forms of gender discrimination—from the academic workplace and from society—dictated the ways that participants would communicate. Laura Miller has dismissed this type of complaint about the Internet, warning that we need to be wary of the media's attempt to cast women as victims on the Internet, as in discussions of flaming or the prevalence of pornography:

> The idea that women merit special protections in an environment as incorporeal as the Net is intimately bound up with the idea that women's minds are weak, fragile, and unsuited to the rough and tumble of public discourse. It's an argument that women should recognize with profound mistrust and resist, especially when we are used as rhetorical pawns in a battle to regulate a rare (if elite) space of gender ambiguity. (1995: 58)

What is needed in this discussion between two different camps of feminists—those who see the Internet as rife with sexism (a group Miller unhelpfully calls the 'schoolmarms') and those who see such behaviour as incidental problems, easily overcome by assertive behaviour—is research that links women's reaction and attraction to the Internet to other aspects of their lived experience, including education, class, sexual orientation, religion, and workplace culture, including how frequently they experience and how they deal with sexual harassment in their workplace.

As Cynthia Cockburn reminds us, 'Gender is a social achievement. Technology too' (1992: 39). The trick for media researchers will be to avoid a kind of static gender essentialism, where women, the elderly, and the poor are eternally, innately deficient *vis-à-vis* new technologies, while all men benefit from inborn technical know-how, an affinity to toying with machines, and an enthusiastic embrace of time-intensive forms of hobbyism. Ethnographic studies can help to provide materialist explanations for women's reluctance to immerse themselves in computer use and on-line communications, as well as explicating the circumstances under which women are attracted to and excel at computer skills.

The role of feminist researchers will include tempering enthusiasm for these new technologies, as Kramarae and Kramer argue:

> The so-called popular media treat the new electronic systems as sexy, but the media fail to deal seriously with most of the gender-related issues. As those who are working closely with its developments know, the Internet will not, contrary to what the media tout, ride the world of hostility, ignorance, racism, sexism, greed and undemocratic governments. The Internet has the potential for creating a cooperative collective international web; but, as with all technologies, the Internet system will be shaped by prevailing communication behaviors, economic policy, and legal decisions. (1995: 15)

One of the most important functions of ethnography, then, may be as an antidote to the hype about computers. Ethnography can describe, for example, the full context in which barriers to entrance onto the information highway exist—especially among people still worrying about keeping a car running on the real highway, much less having the extra time and money to maintain a computer at home.

When turning to the Internet, it will be especially important to treat computer communications as deeply cultural, as employing fiction and fantasy as well as information, and as open to variable interpretations based on gender, race, and ethnicity. As the Internet is championed as the ultimate in 'interactive' media use, it will be extremely important for media researchers to put forward appropriate scepticism about the laudatory uses of the term 'active' in discussions of media use, and to problematize the complex factors involved in attracting both television viewers and computer users to particular contents and genres.

Ethnography can offer an appreciation of some of television's advantages over computers (such as its accessibility by more than one person at a time, and its visual and aural modes of communication) and guard against the unnecessary pathologization of television viewing, which almost always acts to stigmatize the already socially powerless. It will be important to look at the reasons why and the contexts in which television might be more appealing than the Internet. We know that television—for all of its failings—has been cheap, easy to watch at home, enjoyable to talk about with others, and reasonably successful as an educational tool. It will be important to understand the ways in which such social features of broadcast technologies are missing or modified in new technologies. Finally, scholars should work to change the media and the structures of access to communication technologies, while politicizing the public discussion of media in ways that make explicit the gains and the losses at stake in promoting different representations of TV audiences and computer users.

7
Conclusion

You can talk to your wife any time.
All we ask is 8 hours a day.
Books are overrated.
Hobbies. Schmobbies.

Billboards for ABC Television network

ABC's billboards reclaim television viewing with an in-your-face joke at its detractors. The ads take on the social condemnation of TV viewing as a badge of honour, of hipness. Such an invitation to enjoy television despite its negative connotations is perhaps only possible in the age of the Internet. Television audiences are no longer the desirable demographic they once were. Couch potatoes pale in comparison to the targeted niches of superconsumers available through direct marketing campaigns in new forms of digital communication. The sexy high spenders are the Internet surfers. They are guys—and I do mean guys—too busy to watch TV anymore. It is interesting to compare the ABC campaign to a campaign for Cybergold, an Internet advertising service: 'He's worldly. He's influential. We can introduce you to him.'

People are proud of surfing the Net all night, or fiddling endlessly with computers. By comparison, television viewing can be a touchy subject, precisely because of its association with a lack of education, with idleness and unemployment, and its identification as an 'addiction' of women and children. The case studies presented in this book have demonstrated some of the defensiveness that middle-class men and women unprotected by academic credentials may feel in admitting to television viewing because of its connotations of feminine passivity, laziness, and vulgarity. Reading anything is preferable to watching the television. TV is the lowliest of media: even Bourdieu himself has come out with a resounding attack on its worthlessness (Eakin 1997). People always compare their own television viewing to that of the imagined mass audience, one that is more interested, more duped, more entertained, more gullible than they themselves.

Among members of the middle class—those with education but not much economic capital—anxieties about media effects are most acute. I suspect that this group cannot afford the luxury of indulging a love of popular culture, for fear that they will appear uneducated. In interviews, I have found that those with the largest stake in appearing educated make the 'worst' informants about television viewing. When the Montessori teacher claimed, for example, not to know who the popular TV character Barney is, I assumed an element of disingenuousness. Similarly, the accountant father and schoolteacher mother who bemoaned even the watching of *Sesame Street*, and who repeatedly introduced the *MacNeil-Lehrer News Hour* and the NPR news broadcasts, when others in the group were talking about prime-time network TV, made for frustrating interview subjects.

In these encounters, my stake in a version of television viewing as a potentially interesting, resistive, active experience, and my subjects' stake in presenting themselves as discerning media consumers come most sharply into conflict. In the specifics of the research situation it also means that those without such investments are likely to be more loquacious about TV and thus make more valuable subjects, and to hold a lay theory of media effects more in keeping with the predominant lines of thinking within cultural studies.

Rewards accrue to those—such as TV studies academics—who are in a privileged enough position to adopt an interpretation of their viewing as active rather than passive. But the interpretation of viewing as active is not equally available to all: it depends on one's fears for the future, the degree and nature of social aspiration, one's moral or religious judgement of popular media and consumer culture. For the representations of television viewing find their way into the educational system in real ways: punishing those who know too much about TV, or don't know enough to closet their tastes. A studied, conspicuous distance from television is a mark of distinction. A familiar closeness with computers is increasingly a mark of accomplishment, of success. As we move into the digital age, we need to be very sceptical of certain forms of technological determinism: that personal computers are always better than Nintendo or Sega, which are much more widely distributed in African-American and working-class homes; that CD-ROMs are better than videos.

It is not enough to describe and to challenge snobbery about media forms and technologies. Media scholars should also investigate the unfulfilled desires of the audience. If audience studies have proven that audiences are resilient, innovative, flexible, ingenious, that does not mean that they cannot ask for something more, for something different. We can refute the claim that people (only) want what they already have on television. Christine Geraghty has made the argument this way in a review of audience studies: 'Ironically, what seems to be missing in this work which attempts to value what different audience members say is a sense of what claims audiences might legitimately make for different kinds of television' (Geraghty and Lusted 1997: 155). In this book, I have been critical of teachers and parents who adhere to a strict condemnation of the low and the popular, when I see in these beliefs and behaviours a condemnation of working-class people, and often, by implication, of people of colour.

I am troubled, however, by one of the logical—although unintended—extensions of my argument: that the market should run free in the classroom and in the home. One of the future challenges for audience ethnography is to think about ways to shape the research to support protections of public spaces that are also non-commercial, such as schools, exhibitions, libraries—the Internet itself. As John Corner recommends, at this historical moment:

> [Cultural studies] should retain its founding emphasis on the socially differentiated and political character of cultural experience and on the contingency of aesthetic practice, but it should support those policies (some of which are traditionally liberal) seeking to provide all people with the widest possible access to, and means to enjoy, conventionally 'high' academic forms. The rejection of these as irredeemably middle-class is, like the wholesale dismissal of popular culture by elite commentators, a prescriptive narrowing of cultural experience derived from false socio-political assumptions. (1994: 48)

All adults here wish for a media that would cause them less trouble. As Charlotte Brunsdon put it, audiences are used to 'making the best of a bad job' (1989: 126). Christians want less Satanism; teachers want less violence; parents want less encouragement to consume toys and foods. Asking people to envision forms of media that they might want would be a useful addition to many audience research designs: marketers picturesquely call such strategies 'blue sky' research. In the soap opera study, one of our routine questions was what viewers would like to change about soap operas, what would make soap operas better for them. This question was often met with puzzlement. It was difficult for people to answer because they were so resigned to the form's indelibility that it had never occurred to them to think this way. It was hard for them to think of stories or characters that they might prefer—although many mentioned having fewer rich characters—but fewer commercial interruptions was one of the most frequent responses.

One model for this kind of approach is Klaus Bruhn Jensen's form of focus-group interviewing, borrowed from political organizing in Germany in the 1970s, called workshops of the future. He contrasts traditional focus groups to the workshops:

> In terms of their knowledge interests, studies thus tend to approach respondents as objects of predictive control responding to already existing programming. By contrast, respondents might engage each other as subjects in examining forms of mass communication that do not yet exist, thus performing a critique of current media . . . workshops establish a context of reflexivity that is collective and empowered. (1995: 98)

Groups were asked first to list what they dislike about television, then to rephrase these in positive terms, as an expression of what a better kind of television programming and distribution would be like, and finally to develop an action list of places to start to achieve such a utopian goal. Future audience research might try connecting interpretations of texts—decodings—with desired texts, with ideal texts. This would also have the consequence of uniting two different strands of audience research—the encoding–decoding work which is more linked with the project of media criticism, and the domestic context of consumption, more associated with sociology.

Throughout this book I have argued that using a combination of methods, such as observation, diary, and informal conversations, as well as interviews, will produce more nuanced understandings of media consumption. In contrast to the current dominance of postmodern theories, which create a scepticism about any discussion of methodology, I argue, at the risk of sounding old-fashioned, that the most salutary move for audience research would be to concentrate on expanding the repertoire and intensity of qualitative methods. The best designs are those where researchers make contact with informants repeatedly, for as much time as possible, and under as many different circumstances as possible. It will be necessary to use a combination of interview with other methods, such as observation, diary, and informal conversations, especially because computer use, like television viewing, becomes a casual, unconscious activity, deeply ingrained in everyday life. The best example so far has been the work carried out by Roger Silverstone, Eric Hirsch, and David Morley at the Centre for Research on Information and Communication Technologies (CRICT). In this project, the informants were asked repeatedly about the same topics from a variety of different perspectives. Researchers were able to ask informants to compare different or conflicting information they had given on previous occasions, as the study went along. The CRICT study captured how people explain to themselves what is going on in their domestic lives and the ways in which their relationships within and outside the family are mediated through communications technologies. The study gathered time-use diaries, had subjects draw maps of domestic space, charted networks of kin and friends, discussed family photo albums, and considered household budgets (Silverstone *et al.* 1990a). To make the connections between uses of new technologies and old, and to place computers in perspective, it will be important to consider the broad range of media available in the lives of informants and the possible interrelationships among different forms, such as radio, television, newspapers, and the Internet.

Formidable practical constraints arise when using these methodologies. Such obstacles are rooted in the academic institutions that effectively sponsor such research and the particular problems created by the application of ethnographic methods to the study of contemporary media. Too often, audience studies remain unwritten, as drawers full of transcripts and field notes left behind after funding ran out, collaborative teams split up, or researchers hurried on to the next topic before completing the analysis.

If done well, ethnographic research takes a long time. There is extensive time involved in entry into the field situation: working out the logistics of getting in, making it past various gatekeepers, securing the approval of the ever more vigilant human subjects research committees at major universities. The researcher must be patient through the process of securing trust, establishing rapport, understanding the daily routines. The fieldwork itself produces mounds of data, in the form of field notes, tapes, interviews. If ethnography is really done right, it requires that the researcher contextualize media in a wide range of practices and institutions, which have conventionally been studied by sociologists, psychologists, economists, and others. For example, to understand the pre-school teachers I studied, I needed to situate their classrooms in terms of economic understandings of the daycare industry, and pink-collar employment patterns; the education and child psychology literature provided insights into the pedagogical principles and beliefs underlying the teachers' practices; it was also necessary to read widely in the sociology of childhood. Ethnography requires, as Kamala Visweswaran

(1994) reminds us, both fieldwork and homework; trips to the library as well as immersion in the ethnographic scene.

Its time-consuming, labour-intensive nature makes ethnography a method at odds with the transience of media phenomena. It is a certainty that by the time an ethnographic study of television reception is published in journals or academic presses, the media discussed by the informants will be no longer popular, the films forgotten, stars passé, perhaps even the technologies outmoded. As I work on my study of pre-schools, I cannot set foot in a shopping mall without feeling a clutch of anxiety as I notice that the merchandise deriving from the *Power Rangers*—so frequently mentioned by my informants—is reduced further in price and relegated to shelves closer to the back with each visit. What was a peak phenomenon, mentioned constantly in the press and in everyday conversation when I began my work, will be a boring remainder / rerun by the time it is published.

Ethnography yields little in the way of tangible results for quite a while and thus is in conflict with the escalating demands of universities that researchers publish at accelerating rates. Ethnography produces a much slower rate of results compared to textual analysis or theoretical speculation or survey methods, or even less intensive qualitative methods, such as focus groups. Anthropology departments have historically understood the labour-intensive demands of ethnography; if media studies is now maturing, it may also be able to develop the patience to let fully theorized and painstakingly researched ethnography take place.

An important area of improvement would be a greater concern with representativeness and the determinants of sampling. There is a great temptation to generalize from the findings of very small convenience samples. As Geraghty warns, we need to be careful to respect the parameters of the object of study 'in relation to outcomes. A picture of television viewing based on one category—soap opera viewers, viewers in families, fans, children—cannot be used, as has sometimes been the case, to make general comments about television watching as a whole' (Geraghty and Lusted 1997: 155). Researchers need to think at the outset about samples and locations to think strategically about underrepresented groups, those least served by media in its current social / political manifestations.

What case studies can do is chart in-depth pictures of relations between home and work, between professional and familial obligations, between discourses designed to negotiate media consumption and actual practices, between gender ideologies and interpretations. One of the most important perspectives to come out of audience research has been its detailing of the sexual division of labour as it relates to domestic work and leisure time. Despite the increasing theoretical emphasis on the tenuousness of gender identity, the share of domestic labour shouldered by women, and the segregation of women in lower-wage pink-collar jobs, has hardly been reduced since the second wave of feminism. Contemporary media ethnography needs to emphasize historical perspectives on women's employment patterns, on housework, and on childcare, which help to explain contemporary media and patterns of consumption. They can play a role in investigating 'the clusterings of taste-choices and evaluation . . . Just how differentiated access, dispositions, competences and judgments come about in contemporary society through a mix of economic, social and personal factors should be a question on a variety of research agendas' (Corner 1995: 145). I have argued in this book that it is also increasingly necessary to compare media, as we calculate the impact of the Internet and digital communications on television.

I have argued that audience researchers will need to turn their attention to the connections between televisions and computers. Studying the use of computers may force communication scholars to examine the links between public and private, between work and home, more closely and encourage audience researchers to take stock of the role of television outside the home. While domestic studies of television audiences often report rather depressing views of traditional divisions of labour—and the early studies of computer use in the home seem to follow along similar lines—this picture may become more complicated and capture more nuances if we examine the ways in which women use computers in the workplace and how employment uses affect the likelihood that they will become computer enthusiasts (Haddon 1992: 86). We need to engage both in domestic studies of computers and in the study of public uses of television, while examining the connections between office and home use. Both technologies—computers and televisions—need to be studied in both contexts—the domestic and the workplace.

Yet computer technologies, with their planned obsolescence, further exacerbate the problem of dated research. For example, few of the computer models discussed in the research of the 1980s are on the market in those configurations today. Because of the planned obsolescence of so much digital technology, ethnographies of computer communication are doomed to be outmoded. Yet this is an important distinction between ethnography and market research: academics in the field of media and new technologies can cultivate an interest in the recent past, perhaps to take stock of the changes new technologies are bringing to the home and the workplace. It is also important to note that outmoded computer systems live on as hand-me-downs to poor schools in rural and urban areas, where teachers and families often struggle to keep going computer equipment that the yuppie target market can no longer be bothered with. Academics must cultivate an interest in those written off by market researchers—the millions of people who represent demographics not targeted because they are not deemed potentially lucrative, because they are in the wrong neighbourhood, the wrong region of the country, have the wrong income, the wrong age, the wrong number of children, or the wrong colour. We need to be explicit about the stakes involved in our representations of media audiences and computer users, and present our research in forms that are accessible to journalists and policy-makers. The digital revolution is creating huge numbers of 'information have-nots'. With about one-third of US homes yet to subscribe to cable television, the penetration of home computers is unlikely to reach anything like that of television sets. Researching only those who are already working with computers or have computers in the home will skew our understanding. As Dan Schiller usefully reminds us:

> It is only among the better-off—households with $75,000 or more in annual income—that PCs had become routine, with a 60–65% penetration rate . . . Even the experience of the most favorably endowed

> ones in the global political economy shows, therefore, that the level and character of access remained a function of entrenched income inequality. (1997: 5)

Robert McChesney cautions about the catastrophic impact of the Internet on the widening class divide: 'In a class-stratified, commercially oriented society like the United States, cannot the information highway have the effect of simply making it possible for the well-to-do to bypass any contact with the balance of society altogether?' (1996: 117).

The massive investment in a technology so heavily skewed towards benefiting the white middle class must be viewed with scepticism, and convincing arguments for the importance of sustaining funding for television and video production need to be mounted. John Caldwell has noted the logical leap made by many advocates of 'telecomputers' and the information superhighway 'to a vision of the world in which infinite individual needs are met through interactivity and technological responsiveness'. As Caldwell reminds us: 'Most audiences have yet to clamor for the headaches of menus, interactive branching, and nonlinearity. They want Schwarzenegger and production value, and they want it now. Many audiences do not want Nintendo and e-mail either, they want narrative and character' (1995: 348). Television scholars must be prepared to protest the drastic cuts in funding for public television and for independent productions—television sets and VCRs are to be found in nearly every home (and many classrooms); they will continue to be vastly more accessible than computer communications for a very long time, and are probably a better investment for a democratic society. It will be important to do fieldwork close to home, especially with aggrieved communities whose second-rate access to media technologies and whose representational exclusion from the mass media have long and relatively unchanging histories, and for whom the push towards computerization threatens to exacerbate economic hardship and a widening class divide.

University administrators now pressure academics to form 'partnerships' with business (Soley 1995). All of us would do well to make sure that there are clear-cut differences between our research questions and those of market researchers. Bringing to bear on our analysis a full understanding of economic and historical determinants of the situations we study is an important strategy. Maintaining a high degree of consciousness about the markets currently targeted by business, as well as the spread of marketing techniques globally, is another very important strategy. I feel uneasy recalling that during the years when audience researchers were focusing on close readings of the text, the structure of media corporations underwent massive deregulation, conglomeration, and shifts in power. As we move to studying the confluence of television and other technologies, we should strive to incorporate into this work a greater awareness of accompanying shifts in the political economic structure of the media industries.

Ien Ang's recommendations for further ethnographic audience research centre on the global media economy. Despite profound misgivings about the project of audience research as institutionalized in media corporations and in academic institutions, and its close connections to regulative ideologies, Ang emphasizes the suitability of ethnography to the understandings of local media reception in a global, postmodern economy: 'with a greater sense of unequal power relationships across the boards' and 'of the complicated interlockings of autonomy and dependency' (1996: 171). For Ang, the openness of media texts to varied interpretations that has been discovered by audience researchers must also be recognized as a deliberate strategy of capital in a transnational age:

> [C]ritical theory has changed because the structure of the capitalist order has changed. What it has to come to terms with is not the certainty of (and wholesale opposition to) the spread of a culturally coherent capitalist modernity, but the uncertainty brought about by the disturbing incoherence of a globalized capitalist postmodernity, and the mixture of resistance and complicity occurring within it. (1996: 171)

This leads her to a sober reassessment of audience strategies for resistance, and the significance of political economic analysis in light of the domination of originally US media corporations in the opening markets of the Third World and Asia:

> In the end, however, we do need to return to the very substantial Americanness of much of 'global' media, not only in terms of corporate ownership and working principles, but also, more flagrantly, in terms of symbolic content: images, sounds, stories, names. No amount of transformative interpretation will change this. (1996: 172)

Ethnography is the method best able to capture the particular kinds of experience created by the consumption of American media in Australia and Asia, where 'its cultural consequences in particular localities have hardly begun to be understood' (1996: 161).

While my case studies are drawn from fieldwork in the USA, they point to global economic trends. In these Midwestern classrooms, it is easy to detect the presence of corporate sponsors and evidence of the broad reach of marketing in the lives of children. In the Pacific Northwest, the support-group parents feel besieged by Toys-R-Us—just as parents do in Asia and in Europe. The Christian teachers are motivated by the sense that the American mass media is corrupting youth and breaking down social norms—the very feelings that are helping to fuel the popularity of fundamentalist religions around the world. The circumstances of employment in which the daycare-givers find themselves, where salary and benefits are far below their needs, typify the broad trend towards de-skilling of jobs and the creation of an increasing number of low-wage service jobs.

When television is represented as all powerful and all determining, it directs attention away from other important factors, such as schools, wages, housing, transportation, and health care. Thus, gross discrepancies in money and privilege are covered up by offering the negative image of the TV viewer as the determining factor. When more than 20 per cent of children in the United States live in conditions of poverty, it is criminal to promote the belief that television viewing is the primary factor in poor school performance. We must continue to criticize those politicians or experts who insist that if we merely turn off the set, lives and life chances will magically improve.

Anthropologists have recently bemoaned their lack of influence over policy matters, relative to other social science disciplines. The American Anthropological Association has made a series of recommendations in the book *Diagnosing America* (Forman 1991) for ways that ethnography can be brought to engage more centrally with political questions. One strategy—which can easily be adapted to making media audiences studies more relevant to policy-makers—is to focus research on broadly recognized social problems, in their words, 'to understand and act on perceived disorders in the United States'. In media studies this might mean taking up such topics as media consumption in relation to literacy and education, on the role of the media in children's lives, and differential access to new technologies based on gender, race, and ethnicity, on immigrant communities, and on first-time college-goers. A few examples of work that aspires to such relevance are Marie Gillespie's work on how media is used to negotiate the immigrant and exilic experiences (1995), David Buckingham's interviews with parents and children on emotional responses to television (1996)—which includes very specific policy recommendations for systems of rating videos—and John Corner *et al.*'s work on interpretations of fictional and documentary representations on the subject of nuclear energy (1990). Anthropologists insist that there is reason for optimism about the potential effectiveness of ethnographic studies on policy:

> [T]he kinds of questions anthropologists ask about the communities we study (and the kinds of answers we formulate) are distinctly different from those of our sister social sciences. Our emphasis on pluralism, our understanding of culture, our appreciation for the informant's perspective, and our long-standing recognition that policies incongruent with local understandings are doomed to failure, add up to a distinctive perspective that policy makers are ready and willing to hear. (Forman 1991: 301)

For such an interaction to take place, ethnographers need to concentrate on relating local issues to the national and the global, to the cultural, the economic, and the political; to think about the impact of institutions on local circumstances; and to attempt to interact with and speak beyond their academic readerships to include general audiences, politicians, and other professional groups.

After a long period under which quantitative research on the media seemed to have a hold over the imaginations of policy-makers, a more open-minded attitude to qualitative audience research may now be available. The critique of positivism, and the suspicion that all research is guided in some way by the preconceptions of researchers, is a perspective now available in any newspaper. Ethnography's potential acceptance has also been enhanced by the fact that 'soft' qualitative methods—blue-sky research, focus groups, and the like—are now widely accepted in the fields of business and marketing—and political consulting—and routinely reported (often in overly respectful terms) by journalists. Ethnography employs a more accessible language and is thus more persuasive to general readers than other forms of academic discourse. There is a political potential in ethnography's capacity to convince readers 'that what they are reading is an authentic account by someone personally acquainted with how life proceeds in some place, at some time, among some group, is the basis upon which anything else ethnography seeks to do—analyze, explain, amuse, disconcert, celebrate, edify, excuse, astonish, subvert—finally rests' (Geertz 1989: 143). This vivid, narrative quality of media ethnography can serve to distance readers from the normative expectations about media in everyday life, to ratchet down the absurd utopian images of technological consumption that saturate advertising and the popular press, while building awareness of the impoverished circumstances of most of those on the globe. Media studies would do well to take Clifford Geertz's formulation of the goals of research: 'to enlarge the possibility of intelligible discourse between people quite different from one another in interest, outlook, wealth, and power, and yet contained in a world where, tumbled as they are into endless connection, it is increasingly difficult to get out of each other's way' (Geertz 1989: 147).

Having introduced a different representation of the audience produced through domestic ethnography, it is important to move back to the public realms in which these circulate. We need to gauge the stakes involved in our representation of television viewing. Television audience studies should work to change television itself as well as the popular representation of the audience. Audience studies can remove needless anxieties caused by elitism and the overestimation of television's effects, while politicizing the public discussion of media in ways that would demand something more for those audiences who need it most.

Bibliography

Abelman, R. (1990). 'In Conversation: Donald E. Wildmon, American Family Association', in R. Abelman and S. M. Hoover, *Religious Television*, 175–80.

Abelman, R., and Hoover, S. M. (eds.) (1990). *Religious Television: Controversies and Conclusions*. Norwood, NJ: Ablex.

Acker, J. K. (1973). 'Women and Social Stratification: A Case of Intellectual Sexism'. *American Journal of Sociology* 78/4: 936–45.

—— Barry, K., and Esseveld, J. (1991). 'Objectivity and Truth: Problems in Doing Feminist Research,' in M. M. Fonow and J. A. Cook, *Beyond Methodology*, 133–53.

Allen, R. C. (1985). *Speaking of Soap Operas*. Chapel Hill: University of North Carolina Press.

—— (1989). 'Bursting Bubbles: "Soap Opera," Audiences and the Limits of Genre', in E. Seiter *et al.*, *Remote Control*, 16–43.

American Family Association Journal (1996). August.

Anderson, J. A. (1983). 'Television Literacy and the Critical Viewer', in J. Bryant and J. A. Anderson (eds.), *Children's Understanding of Television: Research on Attention and Comprehension*. New York: Academic Press, 297–330.

Ang, I. (1985). *Watching Dallas: Soap Opera and the Melodramatic Imagination*. London: Routledge.

—— (1991). *Desperately Seeking the Audience*. London and New York: Routledge.

—— (1996). *Living Room Wars: Rethinking Media Audiences for a Postmodern World*. London: Routledge.

Ang, I., and Hermes, J. (1993). 'Gender And / In Media Consumption', in I. Ang, *Living Room Wars*, 109–32.

Auletta, K. (1997). 'The Microsoft Provocateur'. *New Yorker*, 12 May: 66–77.

Bacon-Smith, C. (1992). *Enterprising Women: TV Fandom and the Creation of Popular Myth*. Philadelphia: University of Pennsylvania Press.

Bank, D., and Takahashi, D. (1997). 'Microsoft Plans Big Digital TV Push, Stressing Hardware and Programming'. *Wall Street Journal*, 16 April: B6.

Bartkowski, J. P., and Ellison, C. G. (1995). 'Divergent Models of Childrearing in Popular Manuals: Conservative Protestants Vs. the Mainstream Experts'. *Sociology of Religion* 56: 21–34.

Bausinger, H. (1984). 'Media, Technology and Everyday Life'. *Media, Culture and Society* 6/4: 343–51.

Berk, S. F. (1985). *The Gender Factory: The Apportionment of Work in American Households*. New York: Plenum Press.

Bobo, J. (1995). *Black Women as Cultural Readers*. New York: Columbia University Press.

Bobo, J., and Seiter, E. (1991). 'Black Feminism and Media Criticism: *The Women of Brewster Place*'. *Screen* 32/3: 286–302.

Boddy, W. (1990). *Fifties Television: The Industry and Its Critics*. Urbana: University of Illinois Press.

Bourdieu, P. (1984). *Distinction: A Social Critique of the Judgment of Taste*. Trans. R. Nice. Cambridge, MA: Harvard University Press.

Bourdieu, P., and Wacquant, L. J. D. (1992). *An Invitation to Reflexive Sociology*. Chicago, IL: University of Chicago Press.

Brown, K. M. (1994). 'Fundamentalism and the Control of Women', in J. S. Hawley (ed.), *Fundamentalism and Gender*. New York: Oxford University Press, 175–201.

Brown, M. E. (1990). *Television and Women's Culture: The Politics of the Popular*. London: Sage.

Brunsdon, C. (1983). '*Crossroads*: Notes on a Soap Opera', in E. A. Kaplan (ed.), *Regarding Television*, 76–83.

—— (1989). 'Text and audience', in E. Seiter *et al.*, *Remote Control*, 96–115.

—— (1990). 'Problems with quality'. *Screen* 31/1: 67–90.

—— (1993). 'Identity in Feminist Television Criticism'. *Media, Culture and Society*, 15: 309–20.

Brunsdon, C., and Morley, D. (1978). *Everyday Television: 'Nationwide'*. London: British Film Institute.

Brundson, C., D'Acci, J., and Spigel, L. (eds.) (1997). *Feminist Television Criticism: A Reader*. Oxford: Oxford University Press.

Buckingham, D. (1987). *Public Secrets: East Enders and Its Audience*. London: British Film Institute.

—— (1991). 'What Are Words Worth?: Interpreting Children's Talk about Television'. *Cultural Studies* 5/2: 228–45.

—— (1993). *Children Talking Television: The Making of Television Literacy*. London: Falmer Press.

—— (1996). *Moving Images: Children's Emotional Responses to Television*. Manchester: Manchester University Press.

Busfield, J. (1987). 'Parenting and Parenthood', in G. Cohen (ed.), *Social Change and the Life Course*. London: Tavistock.

Bybee, C. R. (1987). 'Uses and Gratifications Research and the Study of Social Change', in D. L. Paletz (ed.), *Political Communication Research: Approaches, Studies, Assessments*. Norwood, NJ: Ablex.

Caldwell, J. (1995). *Televisuality: Style, Crisis, and Authority in American Television*. New Brunswick, NJ: Rutgers University Press.

Cannon, I. W., Higginbotham, E., and Leung, M. L. A. (1991). 'Race and Class Bias in Qualitative Research on Women', in M. M. Fonow and J. A. Cook, *Beyond Methodology*, 107–18.

Carroll, J., and Roof, W. C. (eds.) (1993). *Beyond Establishment: Protestant Identity in a Post-Protestant Age*. Louisville, KY: Westminster / John Knox Press.

Caruso, D. (1996). 'Microsoft Morphs into a Media Company'. *Wired* 4/6: 126–9.

Chambers, J. (1994). 'An Abomination Onto The Lord'. *Harpers*, March: 30.

Clement, A. (1994). 'Computing at Work: Empowering Action by "Low-Level Users"'. *Communications of the ACM* 37/1: 52–65. Also reprinted in R. Kling, *Computerization and Controversy* (1996).

Clifford, J. (1983). 'On Ethnographic Authority'. *Representations* 1/2: 118–46.

Cockburn, C. (1992). 'The Circuit of Technology: Gender, Identity and Power', in R. Silverstone and E. Hirsch, *Consuming Technologies*, 32–47.

Collins, R. (1992). 'Women and the Production of Status Cultures', in M. Lamont and M. Fournier, *Cultivating Differences*, 213–31.

Corner, J. (1994). 'Debating Culture: Quality and Inequality'. *Media, Culture and Society* 16/1: 141–8.

—— (1995). 'Media Studies and the Knowledge Problem'. *Screen* 36/2: 147–59.

Corner, J., Richardson, K., and Fenton, N. (1990). *Nuclear Reactions: Form and Response in 'Public Issue' Television*. London: John Libbey.

Cowan, R. S. (1986). *More Work for Mother*. New York: Basic Books.

Curran, J. (1996). 'The New Revisionism in Mass Communication Research: a Reappraisal', in J. Curran, D. Morley, and V. Walkerdine (eds.), *Cultural Studies and Communications*. London: Edward Arnold, 256–78.

DELIA, J. (1987). 'Communication Research: A History', in C. R. Berger and S. H. Chaffee (eds.), *Handbook of Communication Science*. Newbury Park, CA: Sage, 20–98.

DHOLAKIA, N., and ARNDT, J. (eds.) (1987). *Changing the Course of Marketing: Alternative Paradigms for Widening Marketing Theory*. Greenwich, CT: JAI Press.

DOBSON, J. C. (1994–5). *Focus On The Family Newsletter*. Colorado Springs, Colorado.

DOUGLAS, M. (1988). 'Goods as a System of Communication', in *In the Active Voice*. London: Routledge and Kegan Paul, 20–9.

DYSON, A. H. (1997). *Writing Superheroes: Contemporary Childhood, Popular Culture, and Classroom Literacy*. New York: Teachers' College Press.

EAKIN, E. (1997). 'Bourdieu Unplugged'. *Lingua Franca*, August: 22–3.

ECO, U. (1976). *Theory of Semiotics*. Bloomington: Indiana University Press.

—— (1977). *The Role of the Reader*. Bloomington: Indiana University Press.

ELLINGSEN, M. (1988). *The Evangelical Movement: Growth, Impact, Controversy, Dialog*. Minneapolis: Augsburg Publishing House.

ELLISON, C. G., BARTKOWSKI, J. P., and SEGAL, M. L. (1996). 'Do Conservative Protestants Spank More Often? Further Evidence from the National Survey of Families and Households'. *Social Science Quarterly* 77: 663–73.

ELLISON, C. G., and SHERKAT, D. E. (1993). 'Obedience and Autonomy: Religion and Parental Values Reconsidered'. *Journal for the Scientific Study of Religion* 32: 313–29.

FISKE, J. (1992). 'British Cultural Studies and Television', in R. C. Allen (ed.), *Channels of Discourse Reassembled: Television and Contemporary Criticism*, 2nd edn. Chapel Hill, North Carolina: University of North Carolina Press, 284–326.

FONOW, M. M., and COOK, J. A. (eds.) (1991). *Beyond Methodology: Feminist Scholarship as Lived Research*. Bloomington: Indiana University Press.

FORMAN, S. (1994). *Diagnosing America: Anthropology and Public Engagement*. Ann Arbor: University of Michigan Press.

FURNHAM, A. (1988). *Lay Theories: Everyday Understandings of Problems in the Social Sciences*. Oxford, England: Pergamon.

GARFINKEL, H. (1976). *Studies in Ethnomethodology*. Englewood Cliffs, NJ: Prentice-Hall.

GEERTZ, C. (1988). *Works and Lives: The Anthropologist as Author*. Stanford: Stanford University Press.

Georgia Tech Research Corporation, Graphics, Visualization and Usability Centre (GVU's WWW Surveying Team) www-survey@cc.gatech.edu (1995).

GERAGHTY, C., and LUSTED, D. (eds.) (1997). *The Television Studies Book*. London: Edward Arnold.

GIACQUINTA, J. B., BAUER, J. A., and LEVIN, J. E. (1993). *Beyond Technology's Promise: An Examination of Children's Educational Computing at Home*. Cambridge and New York: Cambridge University Press.

GILBERT, J. (1978). *A Cycle of Outrage*. Chicago, IL: University of Chicago Press.

GILLESPIE, M. (1995). *Television, Ethnicity and Cultural Change*. London and New York: Routledge.

GORDON, J. M. (1991). *Evangelical Spirituality: From the Wesleys to John Stott*. London: SPCK.

GRAY, A. (1987). 'Behind Closed Doors: Video Recorders in the Home', in H. Baehr and G. Dyer (eds.), *Boxed In: Women and Television*. New York: Pandora, 38–54.

—— (1992). *Video Playtime: The Gendering of a Leisure Technology*. London and New York: Routledge.

GRAY, H. (1993). 'The Endless Slide of Difference'. *Critical Studies in Mass Communication* 10: 190–7.

GRIPSRUD, J. (1995). *The Dynasty Years: Hollywood Television and Critical Media Studies*. London: Comedia / Routledge.

Haddon, L. (1992). 'Explaining ICT Consumption: The Case of the Home Computer', in R. Silverstone and E. Hirsch, *Consuming Technologies*, 82–96.

Hall, J. R. (1992). 'The Capital(s) of Cultures: A Nonholistic Approach to Status Situations, Class, Gender, and Ethnicity', in M. Lamont and M. Fournier (eds.), *Cultivating Differences*, 257–88.

Hall, S. (1973). 'Encoding / Decoding in Television Discourse', reprinted in S. Hall *et al.* (eds.), *Culture, Media, Language*. London: Hutchinson, 1981.

Hardacre, H. (1993). 'The Impact of Fundamentalism on Women, the Family and Interpersonal Relations', in M. E. Martin and R. S. Appleby (eds.), *Fundamentalism and Society: Reclaiming the Sciences, the Family, and Education*. Chicago, IL: University of Chicago Press, 124–47.

Harding, S. (ed.) (1987). *Feminism and Methodology: Social Science Issues*. Bloomington: Indiana University Press.

Helm, L. (1996). 'Microsoft Unveils Revamped Online Service'. *Los Angeles Times*, 11 October: D2.

Hendershot, H. (1995). 'Shake, Rattle and Roll: Production and Consumption of Fundamentalist Youth Cultures'. *Afterimage* 22/7: 19–22.

Hermes, J. (1995). *Reading Women's Magazines: An Analysis of Everyday Media Use*. Cambridge, UK: Polity Press.

Herring, S. C. (1996). 'Gender and Democracy in Computer-Mediated Communication'. *Electronic Journal of Communication* 3/2. Reprinted in S. C. Herring, *Computer-Mediated Communication: Linguistic and Cultural Perspectives*. Amsterdam, Philadelphia: J. Benjamins (1996), 225–46.

Hobson, D. (1982). *'Crossroads': Drama of a Soap Opera*. London: Methuen.

—— (1989). 'Soap Operas at Work', in E. Seiter *et al.*, *Remote Control*, 150–68.

Hochschild, A. (1989). *The Second Shift*. New York: Avon Books.

—— (1997). *The Time Bind: When Work Becomes Home and Home Becomes Work*. New York: Metropolitan Books.

Hodas, S. (1993). 'Technology Refusal and the Organizational Culture of Schools'. *Electronic Journal of Education Policy Analysis Archives*, 1/10. Reprinted in R. Kling, *Computerization and Controversy* (1996), 197–218.

Hodge, B., and Tripp, D. (1986). *Children and Television: A Semiotic Approach*. Palo Alto: Stanford University Press.

Hollway, W. (1989). *Subjectivity and Method in Psychology: Gender, Meaning, Science*. London: Sage.

Hood, P. (1997). 'The Wizard of Silicon Valley'. *Newmedia* 6: 14.

Hoover, S. M. (1988). *Mass Media Religion: The Social Sources of the Electronic Church*. Newbury Park: Sage.

Hunter, J. D. (1987). *Evangelicalism*. Chicago, IL: University of Chicago Press.

Iacono, S., and Kling, R. (1996). 'Computerization Movements and Tales of Technological Utopianism', in R. Kling, *Computerization and Controversy*, 85–107.

Indiana Association for the Education of Young Children (IAEYC) (1993). *Worthy Wage Campaign*.

Ireland, C., and Johnson, B. (1995). 'Exploring the Future in the Present'. *Design Management Journal* 6/2: 57–64.

Iser, W. (1978). *The Act of Reading: A Theory of Aesthetic Response*. Baltimore, MD: John Hopkins Press.

Jenkins, H. (1992). *Textual Poachers: Television Fans and Participatory Culture*. London and New York: Routledge.

Jensen, K. B. (1995). *The Social Semiotics of Mass Communication*. London: Sage.

Jhally, S., and Lewis, J. (1992). *Enlightened Racism: 'The Cosby Show', Audiences, and the Myth of the American Dream*. Boulder, CO: Westview Press.

Kaplan, E. A. (ed.) (1983). *Regarding Television: Critical Approaches—An Anthology*. Frederick, MD: University Publications of America.

Kelley, D. (1972). *Why Conservative Churches are Growing*. New York: Harper and Row.

Kintz, L. (1997). *Between Jesus and the Market: Emotions That Matter in Right-Wing America*. Durham, NC: Duke University Press.

Kintz, L., and Lesage, J. (eds.) (1998). *Culture, Media and the Religious Right*. Minneapolis: University of Minnesota Press.

Klatch, R. (1987). *Women and the New Right*. Philadelphia: Temple University Press.

Kleinman, S., and Coop, M. A. (1993). *Emotions and Fieldwork*. Newbury Park, CA: Sage Publications.

Kling, R. (ed.) (1996). *Computerization and Controversy: Value Conflicts and Social Choices*, 2nd edn. San Diego: Academic Press.

Kramarae, C. (ed.) (1988). *Technology and Women's Voices: Keeping in Touch*. New York: Routledge and Kegan Paul.

Kramarae, C., and Kramer, J. (1995). 'Legal Snarls for Women in Cyberspace'. *Internet Research: Electronic Networking Applications and Policy* 5: 14–24.

Lamont, M., and Fournier, M. (eds.) (1992). *Cultivating Differences: Symbolic Boundaries and the Making of Inequality*. Chicago, IL: University of Chicago Press.

Lamont, M., and Lareau, A. (1988). 'Cultural Capital: Allusions and Glissandos in Recent Theoretical Developments'. *Sociological Theory* 6: 163.

Lawless, E. (1988). *Handmaidens of The Lord: Pentecostal Women Preachers and Traditional Religion*. Philadelphia: University of Pennsylvania Press.

Lemish, D. (1987). 'Viewers in Diapers: The Early Development of Television Viewing', in T. R. Lindlof (ed.), *Natural Audiences*, 33–57.

Lewis, J. (1991). *The Ideological Octopus: An Exploration of Television and Its Audience*. New York: Routledge.

Lewis, L. (ed.) (1992). *The Adoring Audience: Fan Culture and Popular Media*. London and New York: Routledge.

Lesch, S. G. (1994). 'Your Work On Television? A View from the USA'. *Computer-Mediated Communication Magazine* 1/4: 5.

Liebes, T. (1990). *The Export of Meaning: Cross-Cultural Readings of 'Dallas'*. New York: Oxford University Press.

Lienesch, M. (1993). *Redeeming America: Piety and Politics in the New Christian Right*. Chapel Hill: University of North Carolina.

Lindlof, T. R. (ed.) (1987). *Natural Audiences: Qualitative Research of Media Uses and Effects*. Norwood, NJ: Ablex.

—— (1995). *Qualitative Communication Research Methods*. Thousand Oaks, CA: Sage Publications.

Lipsitz, G. (1990). *Time Passages: Collective Memory and American Popular Culture*. Minneapolis: University of Minnesota Press.

Livingstone, S. M. (1990). *Making Sense of Television: The Psychology of Audience Interpretation*. Oxford, England and New York: Pergamon Press.

Livingstone, S. M. (1992). 'The Meaning of Domestic Technologies: A Personal Construct Analysis of Familial Gender Relations', in R. Silverstone and E. Hirsch, *Consuming Technologies*, 113–30.

Livingstone, S. M., and Lunt, P. (1994). *Talk on Television: Audience Participation and Public Debate*. New York: Routledge.

Lull, J. (ed.) (1988). *World Families Watch Television*. Newbury Park, CA: Sage.

—— (1991). *China Turned On: Television, Reform, and Resistance*. London: Routledge.

Marsden, G. (1980). *Fundamentalism and American Culture: The Shaping of Twentieth Century Evangelicalism*. New York: Oxford University Press.

Marcus, G. E., and Fischer, M. J. (1986). *Anthropology as Cultural Critique*. Chicago, IL: University of Chicago Press.

McChesney, R. W. (1996). 'The Internet and U.S. Communication Policy-Making in Historical and Critical Perspective'. *Journal of Communication* 46/1: 98–124.

McRobbie, A. (1991). *Feminism and Youth Culture: From 'Jackie' to 'Just Seventeen'*. Boston: Unwin Hyman.

Miller, L. (1995). 'Women and Children First: Gender and the Settling of the Electronic Frontier', in J. Brook and I. A. Boal (eds.), *Resisting The Virtual Life*. San Francisco: City Lights.

Miner, B. (1996). 'Splits on the Right: What Do They Mean for Education?' *Rethinking Schools*, Spring: 10–15.

Modleski, T. (1984). *Loving with a Vengeance: Mass-Produced Fantasies for Women*. New York: Methuen.

Moores, S. (1993). *Interpreting Audiences: The Ethnography of Media Consumption*. London: Sage.

Morley, D. (1980). *The 'Nationwide' Audience*. London: British Film Institute.

—— (1986). *Family Television*. London: Comedia / Routledge.

—— (1992). *Television, Audiences and Cultural Studies*. London: Routledge.

Morris, M., and Ogan, C. (1996). 'The Internet as a Mass Medium'. *Journal of Communication* 46/1: 39–50.

Murdock, G., Hartmann, P., and Gray, P. (1992). 'Contextualizing Home Computing: Resources and Practices', in R. Silverstone and E. Hirsch, *Consuming Technologies*, 146–60.

Nelson, M. K. (1990). 'Mothering Others' Children: The Experiences of Family Day Care Providers', in E. K. Abel and M. K. Nelson (eds.), *Circles of Care: Work and Identity in Women's Lives*. Albany, NY: State University of New York Press, 210–33.

Newhagen, J., and Rafaeli, S. (1996). 'Why Communication Researchers Should Study the Internet: A Dialogue'. *Journal of Communication* 46/1: 4–14.

Nightingale, V. (1996). *Studying Audiences: The Shock of the Real*. London: Routledge.

Palmer, P. (1986). *The Lively Audience: A Study of Children Around the TV Set*. Sydney: Unwin Hyman.

Pettey, G. R. (1990). 'Bibles, Ballots and Beatific Vision: The Cycle of Religious Activism in the 1980s', in R. Abelman and S. M. Hoover, *Religious Television*, 197–208.

Press, A. (1991). *Women Watching Television*. Philadelphia: University of Pennsylvania.

Press, A. L., and Cole, E. R. (1995). 'Reconciling Faith and Fact: Pro-Life Women Discuss Media, Science and the Abortion Debate'. *Critical Studies in Mass Communication* 12/4: 380–402.

—— —— (1998). *Imagining Our Lives: Television, Women's Talk and the Political Culture of Abortion*. Chicago, IL: University of Chicago Press.

Quebedeaux, R. (1976). *The New Charismatics: The Origins, Development, and Significance of Neo-Pentecostalism*. New York: Doubleday.

Radway, J. (1984). *Reading the Romance: Women, Patriarchy and Popular Literature*. Chapel Hill: University of North Carolina Press.

Rakow, L. (1988). 'Women and the Telephone: The Gendering of a Communications Technology', in C. Kramarae (ed.), *Technology and Women's Voices*, 207–28.

Rubin, S. (1996). 'Conservative Spotlight: American Family Association'. *Human Events* 5/April: 6–7.

Schiller, D. (1997). 'Les marchands à l'assaut de l'Internet' (Cornering the Market in Cyberspace). *Le Monde Diplomatique* 516/March: 1, 24, 25.

Schlesinger, P., Dobash, R. E., Dobash, R. P., and Weaver, C. K. (1992). *Women Viewing Violence*. London: British Film Institute.

Schwichtenberg, C. (1994). 'Gender and media audiences', in J. Cruz and J. Lewis (eds.), *Viewing, Reading, Listening: Audiences and Cultural Reception*. Boulder: Westview Press, 133–48.

Seiter, E. (1990). 'Making Distinctions in Audience Research'. *Cultural Studies* 4/1: 61–84.

—— (1993). *Sold Separately: Children and Parents in Consumer Culture*. New Brunswick, NJ: Rutgers University Press.

—— (1996). 'Mothers Watching Children Watching Television', in B. Skeggs (ed.), *Feminists Produce Cultural Theory*. Manchester: Manchester University Press.

Seiter, E., Borchers, H., Kreutzner, G., and Warth, E. (eds.) (1989). *Remote Control: Television, Audiences and Cultural Power*. London: Routledge.

Shattuc, J. (1997). *The Talking Cure: TV Talk Shows and Women*. New York: Routledge.

Shumate, R. (1996). 'The Gospel of Power'. *Eastbay Out Now*, April 9–22: 7.

Silverstone, R. (1991). 'Beneath the Bottom Line: Households and Information and Communication Technologies in an Age of the Consumer'. London: PICT (Programme in Information and Communication Technologies), Policy Research Papers No. 17.

—— (1994). *Television and Everyday Life*. London: Routledge.

Silverstone, R., and Hirsch, E. (eds.) (1992). *Consuming Technologies: Media and Information in Domestic Spaces*. London: Routledge.

Silverstone, R., Hirsch, E., and Morley, D. (1990a). 'Information and Communication Technologies and the Moral Economy of the Household'. CRICT Discussion Paper, Brunel University. Reprinted in R. Silverstone and E. Hirsch, *Consuming Technologies* (1992).

—— —— —— (1990b). 'Listening to a Long Conversation: An Ethnographic Approach to the Study of Information and Communication Technologies in the Home'. CRICT Discussion Paper, Brunel University. Reprinted in *Cultural Studies* 5/2 (1991).

Skeggs, B. (ed.) (1996). *Feminist Cultural Theory: Process and Production*. Manchester and New York: Manchester University Press.

—— (1997). *Formations of Class and Gender: Becoming Respectable*. London: Sage.

Smith, D. E. (1987). *The Everyday World as Problematic: A Feminist Sociology*. Boston: Northeastern University Press.

—— (1990a). *The Conceptual Practices of Power: A Feminist Sociology of Knowledge*. Boston: Northeastern University Press.

—— (1990b). *Texts, Facts, and Femininity: Exploring the Relations of Ruling*. London and New York: Routledge.

Soley, L. (1995). *Leasing the Ivory Tower: The Corporate Takeover of Academia*. Boston: South End Press.

Spender, D. (1995). *Nattering on the Net: Women, Power and Cyberspace*. North Melbourne, Australia: Spinifex Press.

Spigel, L. (1992). *Make Room for TV: Television and the Family Ideal in Postwar America*. Chicago, IL: University of Chicago Press.

—— (1996). 'Portable TV: Studies in Domestic Space Travel', paper delivered at Console-Ing Passions: The Annual Conference on Television, Video, and Feminism.

Stacey, J. (1988). 'Can There Be a Feminist Ethnography?' *Women's Studies International Forum* 11/1: 21–7.

—— (1991). *Brave New Families*. New York: Basic Books.

Star, S. L. (1995). 'Introduction', *The Cultures of Computing*. Oxford: Blackwell, 1–28.

Strasser, S. (1982). *Never Done: A History of American Housework*. New York: Pantheon Books.

Taylor, H. J., Kramarae, C., and Ebben, M. (eds.) (1993). *Women, Information Technology and Scholarship*. Urbana, IL: Women and Information Technology and Scholarship Colloquium, Center for Advanced Study, University of Illinois.

Thomas, L. (1995). 'In Love with Inspector Morse: Feminist Subculture and Quality Television'. *Feminist Review* 51: 1–25.

Tulloch, J. (1990). *Television Drama: Agency, Audience, Myth*. London and New York: Routledge.

Turkle, S. (1995). *Life on the Screen*. New York: Simon and Schuster.

van Zoonen, L. (1994). *Feminist Media Studies*. London: Sage.

Visweswaran, K. (1994). *Fictions of Feminist Ethnography*. Minneapolis: University of Minnesota Press.

Walkerdine, V. (1990). *School Girl Fictions*. London and New York: Verso.

—— (1993). ' "Daddy's Gonna Buy You a Dream to Cling to (And Mummy's Gonna Love You Just as Much as She Can)": Young Girls and Popular Television', in D. Buckingham (ed.), *Reading Audiences: Young People and The Media*. Manchester and New York: Manchester University Press, 74–88.

Wheelock, J. (1992). 'Personal Computers, Gender and an Institutional Model of the Household', in R. Silverstone and E. Hirsch, *Consuming Technologies*, 97–112.

Williams, R. (1974). *Television: Technology and Cultural Form*. London: Fontana.

Wrigley, J. (1990). 'Children's Caregivers and Ideologies of Parental Inadequacy', in E. K. Abel and M. K. Nelson (eds.), *Circles of Care: Work and Identity in Women's Lives*. Albany, NY: State University of New York Press, 283–308.

Index